Praise for
*Why Science Does Not Disprove God*

"Aczel takes aim at the New Atheists in his intelligent and stimulating book. . . . Part of the continuing and restorative conversation of humanity with itself. In the end, all of our art, our science, and our theological beliefs are an attempt to make sense of this fabulous and fleeting existence we find ourselves in."
—*Washington Post*

"Aczel's book goes a long way to showing why many scientists and philosophers misunderstand each other when it comes to how to approach ultimate questions. . . . Well worth a read."
—*Forbes*

"If everyone understood as well as Amir Aczel does that scientific and religious ways of knowing belong to entirely separate and uncompeting forms of human experience, the world would be a much more pleasant place to live in."
—Ian Tattersall, American Museum of Natural History (Division of Anthropology); author of *Masters of the Planet: The Search for Our Human Origins*

"Amir Aczel combines scientific credibility, stylistic elegance, and argumentative vigor in *Why Science Does Not Disprove God*. . . . What's more, he's right."
—Rabbi David Wolpe, Sinai Temple, Los Angeles; author of *Why Faith M*---

"In this interesting book, Amir Aczel does not intend to *prove* the existence of God, and in fact, he admits that 'the God of literal interpretations of Scripture . . . certainly does not exist.' Rather, he argues that science *has not disproved* the existence of God. Whether you are persuaded by his reasoning or not, you will greatly enjoy this thoughtful, erudite journey through modern science and philosophy, and the clear exposition of a problem with which humans have struggled for millennia."

—Mario Livio, astrophysicist, Space Telescope Science Institute; author of *Is God a Mathematician?* and *Brilliant Blunders*

"Well-informed and readable." —*Wall Street Journal*

"In Aczel, Richard Dawkins and his fellow New Atheists face a formidable opponent. As a mathematician with an impressive Berkeley-Harvard résumé, Aczel wields impressive intellectual weapons in demolishing the New Atheists' claims that science has disproven the existence of God. With compelling reasoning, Aczel demonstrates that whenever Dawkins and his allies turn their attacks against anything but naively literal readings of the Bible, they distort or misrepresent the methods and findings of science." —*Booklist* (starred review)

"Aczel is one of our best science popularizers."

—*Publishers Weekly*

"Amir Aczel is a pop idol of the science-writing world."

—*Willamette Week*

# WHY
# SCIENCE
# DOES NOT
# DISPROVE
# GOD

Also by Amir D. Aczel

# WHY SCIENCE DOES NOT DISPROVE GOD

## AMIR D. ACZEL

WILLIAM MORROW

*An Imprint of HarperCollinsPublishers*

*For Debra*

WHY SCIENCE DOES NOT DISPROVE GOD. Copyright © 2014 by Amir Aczel. All rights reserved. Printed in the United States of America. No part of this book may be used or reproduced in any manner whatsoever without written permission except in the case of brief quotations embodied in critical articles and reviews. For information address HarperCollins Publishers, 195 Broadway, New York, NY 10007.

HarperCollins books may be purchased for educational, business, or sales promotional use. For information please e-mail the Special Markets Department at SPsales@harpercollins.com.

A hardcover edition of this book was published in 2014 by William Morrow, an imprint of HarperCollins Publishers.

FIRST WILLIAM MORROW PAPERBACK EDITION PUBLISHED 2015.

*Designed by Diahann Sturge*

Library of Congress Cataloging-in-Publication Data has been applied for.

ISBN 978-0-06-223060-7

15 16 17 18 19   OV/RRD   10 9 8 7 6 5 4 3 2 1

# Contents

# Introduction

In a televised debate about religion and science held at La Ciudad de las Ideas, the international conference of ideas in Puebla, Mexico, in November 2010, the prominent British evolutionary biologist and atheist Richard Dawkins took a novel tack: he argued that our understanding of *physics* is the new major source of proof that any assumption of a "creator" is unnecessary. By then, Dawkins had been using biology and evolution to argue against the existence of God for many years. I was there on the stage with Dawkins in Mexico, and the experience led me to the thesis of this book: that science has not provided any proof that the existence of a divine creator of some kind must necessarily be false. And in the chapters that follow I will show that we have by no means reached the point at which people can claim, in the name of science, that God does not exist.

While I was listening to Dawkins misuse concepts from mathematics and physics, episodes from my life as a professor of mathematics and statistics, as well as a science writer, flashed through my mind. As an undergraduate at the University of California at Berkeley in the 1970s, I worked in the lab of Professor Gabor Somorjai, a physical chemist who was hoping to uncover the secret of life through processes taking place on a platinum crystal lattice. His theory was that catalysis on a primeval crystal surface hundreds of millions of years ago led to the evolution of life. Somorjai failed. Hard as he tried, he could never replicate the mysterious processes that produce life from inanimate chemical elements and compounds; his lack of success humbled him, and gave him a renewed sense of wonder about the universe. (He still made his name by contributing to the invention of the catalytic converter used in cars.)

I also remembered learning the immensely complicated and surreal laws of quantum mechanics—first in an extraordinary lecture given at Berkeley by one of the greatest quantum pioneers, Werner Heisenberg, just a few years before he died, and later in advanced physics classes. These laws are so incomprehensible to us—as Richard Feynman said, "If you think you understand quantum mechanics, then you don't understand quantum mechanics"—that I was baffled to find that Dawkins and a number of cosmologists were trying to definitively argue that these weird quantum rules bring about a universe "out of nothing," and that therefore there is no God.

As a science writer concentrating on mathematics, physics, and cosmology, I have marveled myriad times about what to me is one of the greatest mysteries of all: how, within the immensely hot and dense "soup of particles" that constituted our universe, a fraction of a second after the Big Bang, the quarks suddenly gathered in threes: two "ups" and a "down" to form protons and two "downs" and an "up" to form neutrons. How was it ever possible, I have asked myself, that the charges of these quarks turned out to be exactly ⅔ for an "up" and –⅓ for a "down", so that the proton would miraculously match the opposite charge of the electron (–1) and the neutron's charge would be precisely zero? How did such an incredibly improbable event ever happen without some calculated act of creation? And further, how did the masses of the elementary particles turn out to have the perfectly precise ratios needed so that our world of atoms and molecules could exist at all? How did the forces of nature—gravity, electromagnetism, and the weak and strong nuclear forces acting inside nuclei, as well as the mysterious "dark energy" that permeates space—receive just the right strengths they need to maintain a universe that has the required stability and neither collapses onto itself nor explodes before life has a chance to evolve? It is hard to imagine all this happening just by chance.

Equally, having conducted a major study of consciousness and how it may have arisen in early hominids for my book on Peking Man, *The Jesuit and the Skull*, for which I interviewed some of the world's leading anthropologists and prehistoric archaeologists, I have learned how much we do not know about consciousness, what it means, what it signifies as a stage of human development, and how it came about. In short, there's a great deal we have yet to understand even about *ourselves*, much less God.

We still have little idea how the complicated, eukaryotic cells in the bodies of living organisms emerged, and with them the complex life-forms we see on Earth. In science, the fine-tuning of the parameters required for life has such an incredibly small probability to have arisen that the famous British cosmologist Stephen Hawking has described it as follows: "If one considers the possible constants and laws that could have emerged, the odds against a universe that has produced life like ours are immense." Another leading cosmologist, Roger Penrose, addressing only *one* of the many parameters necessary for a universe that would support life, has put the probability of the emergence of our universe as 1 over $10^{10^{123}}$, meaning 1 divided by 10 raised to the power of 10, raised to the power of 123. Such numbers are humbling. Now consider the odds of intelligent life developing. To decisively assert that there was no God or act of creation behind our immeasurably unlikely universe seems presumptuous.

**SO THE IDEA** of writing a book stating my conviction that modern science has not disproved the existence of God arose in my mind during the debate at La Ciudad de las Ideas, and later grew during three further annual sessions of this stimulating congress.

The Puebla conference, which features some of the world's leading thinkers, writers, and academics, is the brainchild of a gifted individual: Dr. Andres Roemer, a Mexican intellectual and television personality. Soon after I was awarded a Guggenheim Fellowship in 2005, Dr. Roemer contacted me through the Guggenheim Foundation and asked if I would be interested in participating in his proposed first conference. We met in Cambridge, Massachusetts, and I immediately offered my help and participation. My debate with Dawkins was a small part of this larger enterprise.

The past few years have seen the rapid growth of the idea that God and science cannot possibly coexist. I feel that many people who hold this view distort both the process of science and its value. Science is about the objective pursuit of truth, and we should be very skeptical when "science" is invoked to further someone's sociocultural agenda.

*The purpose of this book is to defend the integrity of science.*

I firmly believe that spirituality, religion, and faith have important roles to play in our lives. They keep us humble in the face of the great wonders of nature; they help maintain our social values, promoting the care for our weak and poor; and they provide hope and some moral code in our ever more complicated modern world. Science and spirituality are both integral parts of the human search for truth and meaning; they provide us possible paths of comprehending and appreciating the vast cosmos and our place in it.

This book is not written from the perspective of any one faith tradition, nor does it seek to defend our often flawed religious institutions. "God" here is used in the broadest possible sense: the Creator. Spirituality, including religious faith, is understood to be the age-old human impulse to know, respond, and possibly align with this absolute, original, and eternal force, without which the universe would not exist. My goal is to restore science and faith to their proper realms and end the confusion sown by those who aim to destroy faith in the name of science.

As a science writer who has made a career of reporting on some of the most complicated and exciting advances in science and mathematics over the past quarter century, I realize that in publishing this book I am taking a risk. In these pages I am attacking the arguments of many prominent scientists and thinkers, and I understand that my doing so will likely lead to criticisms of my book. But I feel very strongly that the integrity of science has been compromised by some New Atheist writers and that it is important to set things straight and to restore the distinction between rigorous logic and overreaching supposition.

**I THANK VERY** warmly and affectionately my good friend Andres Roemer for his repeated invitations to me to participate in his Ciudad de las Ideas conferences, and to the organizers of the meetings for their hospitality. I also thank the people listed below, with whom I have had the good fortune over the years to meet and discuss science and religion. I am immensely indebted to my literary agent, Albert Zuckerman, and to his dedicated staff at Writers House, for their work in supporting the project of writing this book. I am very grateful to Peter Hubbard, my editor at HarperCollins, for all his efforts and for believing in this book; thanks also to editorial assistant Cole Hager, who was instrumental in making this project run smoothly. I thank the copy editor, Greg Villepique, for his attention to detail, as well as designer Diahann Sturge for help in producing this book.

## Interviews and Discussions

Philip Anderson, Nobel laureate physicist, Princeton University

Alain Aspect, quantum theorist, University of Paris

Ofer Bar-Yosef, archaeologist, Harvard University

Shmuley Boteach, rabbi and author

Deepak Chopra, spirituality author and speaker

Richard Dawkins, evolutionary biologist, Oxford University

Dinesh D'Souza, author on religious and political topics

Jerome I. Friedman, Nobel laureate physicist, Massachusetts Institute of Technology

Sheldon Glashow, Nobel laureate physicist, Boston University

Alan Guth, cosmologist and physicist, Massachusetts Institute of Technology

Akihiro Kanamori, mathematician, Boston University

Thomas King, Jesuit theologian and evolutionist (deceased), Georgetown University

Robert Kurzban, evolutionary psychologist, University of Pennsylvania

Leon Lederman, Nobel laureate physicist who coined the term *God particle,* University of Chicago

Benoit Mandelbrot, mathematician and inventor of fractal geometry (deceased), IBM and Yale University

Gary Marcus, cognitive psychologist, New York University

Roger Penrose, mathematician and physicist, Oxford University

Martin Perl, Nobel laureate physicist, Stanford University

Saul Perlmutter, Nobel laureate astrophysicist, University of California, Berkeley

Thomas Reddy, Jesuit theologian, Vatican City

Martin Rees, Astronomer Royal of the United Kingdom, Cambridge University

Saharon Shelah, mathematician, Hebrew University

Michael Shermer, editor in chief of *Skeptic* magazine

Abner Shimony, physicist and philosopher, Boston University

George Smoot, Nobel laureate physicist, University of California, Berkeley

Jack Steinberger, Nobel laureate physicist, CERN, Geneva

Paul Steinhardt, cosmologist, Princeton University

Leonard Susskind, physicist and cosmologist, Stanford University

Ian Tattersall, paleoanthropologist, American Museum of Natural History

Gerard 't Hooft, Nobel laureate physicist, University of Utrecht

Steven Weinberg, Nobel laureate physicist, University of Texas

John Archibald Wheeler, physicist (deceased), Princeton University

Frank Wilczek, Nobel laureate physicist, Massachusetts Institute of Technology

David Wolpe, rabbi and theologian

Anton Zeilinger, quantum physicist, University of Vienna

# Prologue

## The Birth of the New Atheism

The "New Atheism" movement was launched as a direct consequence of the attacks of September 11, 2001. The coldblooded murder of thousands of unsuspecting innocent people on a beautiful late summer day horrified America and the world. These barbarous acts made many people justly regard with anger a religion that could even hint that its followers should commit such crimes in the name of God. How could any religion, supposedly based on the word of God, lead people to indiscriminately destroy the lives of so many human beings (including, inevitably, some of their coreligionists)? This was the question on many people's minds.

One of the first writers to address these painful issues was Sam Harris, who within three years of the attacks published a book called *The End of Faith: Religion, Terror, and the Future of Reason* (2004). Harris lambasted religion and questioned its value in our lives through the perspective of September 11. He argued passionately that faith—indeed, organized religion of any kind—has no place in the modern world, and that it brings only evil and destruction. And he stated that we may never be able to counter extremist Islam as long as we continue to hold on to our own religious beliefs. Militant religion, according to Harris, can only be fought against rationally and effectively by people who have shed their own faith. He later answered criticisms of his book, which had become an instant best seller but had also generated controversy, in a second popular work, *Letter to a Christian Nation* (2006).

Though he is not a working scientist, Harris used scientific concepts to argue against religion. But the move from the justified moral outrage about 9/11 to a "scientific" argument against faith in general results in a somewhat condescending tone. Here is a sample of what he writes about science:

> All complex life on earth has developed from simpler life-forms over billions of years. This is a fact that no longer admits of intelligent dispute. If you doubt that human beings evolved from prior species, you may as well doubt that the sun is a star. Granted, the sun doesn't seem like an ordinary star, but we know that it is a star, and we know that it is a star that just happens to be relatively close to the earth.

Harris presents here some of the strong points of science: most people today accept evolution as the mechanism for biological change. But his statements that the sun "doesn't seem like an ordinary star" and that it's "a star that just happens to be relatively close to the earth" imply that religious people are as uninformed as children about science.

Harris's argument against religion widens to all areas of modern life:

> One of the most pernicious effects of religion is that it tends to divorce morality from the reality of human and animal suffering. Religion allows people to imagine that their concerns are moral when they are not—that is, when they have nothing to do with suffering and its alleviation.

Unfortunately this notion is typical of the New Atheism. New Atheists argue that there is no connection between religion and morality, often using examples of extreme cruelty such as rapes and murders of young children, as well as genocides, tortures, and other heinous crimes. Harris singles out religious people, and says that their religion contributes to their failure to stop such atrocities. Hence, according to Harris, morality and religion do not overlap. This argument is spurious, however, because nonreligious people have done no better at stopping atrocities. Also, many important charities worldwide are based on religious giving, and to dismiss or ignore them and their influence on human welfare is wrong.

The same year that Harris published *Letter to a Christian Nation*, Richard Dawkins, who for decades had been advocating atheism in public lectures and articles, released a book that received even wider circulation and global acclaim, titled *The God Delusion* (2006). In this work, Dawkins uses his prominence in the field of biology to launch a scientific argument against the existence of God. But in addition to using ideas from science—mainly evolutionary biology, but also a smattering of notions plucked from physics and cosmology—Dawkins embarks on a personal crusade against Scripture, especially the Old Testament.

Quoting passages from the Bible that portray the "Abrahamic God," as Dawkins calls him, as vindictive, cruel, unpredictable, and jealous, he goes on to label God a "psychotic delinquent." Dawkins expresses shock that any thinking human being could ever believe in this flawed God, and then decries aspects of the New Testament as well, questioning how any rational person could accept the notions of virgin birth, resurrection, and other "miracles." Virgin births do not occur in nature, the biologist tells us, and neither do the dead return to life or ascend to heaven. Revelation is not a method of obtaining information about the world, he says: scientific evidence is. On this point, I would certainly agree with Dawkins.

But strangely, Dawkins avoids criticizing both his native Anglicanism (even offering mild praise for some of its clergymen) and all Eastern religions—conveniently identifying them as "ways of life," rather than proper religions.

Along the way, Dawkins shares some of his unscientific beliefs, such as that Hitler was not nearly as evil as Caligula (how does he know?) and that abusing children sexually is not as bad as indoctrinating them in a religion. With respect to this last assertion, he claims to be speaking from personal experience, saying that he was abused as a child but that it amounted to only an "embarrassment," while being exposed to religious ideas caused far more damage. One wonders if the many adults now coming forward with revelations of having been raped or molested as children would agree with this view.

Dawkins's thrust throughout his book is that religion is not only bad but also stupid. Religious people deserve no respect from the rest of society, he says. (In chapter 1 of *The God Delusion*, Dawkins has an entire section titled "Undeserved Respect," referring to his view that religious people are not deserving of respect for their beliefs.) Yet after his impassioned, all-out attack on the folly of religion, Dawkins rates himself only a 6 on an atheism scale he has designed in which 1 stands for absolute belief in God and 7 indicates a 100 percent sureness that there is no God. Why does the world's most prominent atheist suddenly hedge?

In any event, the main purpose of Dawkins's book is to attempt to use science to prove that religion is false and that God does not exist. Dawkins wants to replace "God" with "evolution," and to show that the factors of evolution—survival of the fittest, adaptation to the environment, and natural selection (preferential sexual reproduction for better-adapted individuals)—lead directly to the complexity of life we see around us and that, therefore, there was never a need for any external "creator."

He also argues, albeit briefly and haphazardly, that the universe as a whole came about through a nonbiological mechanism he claims must be "similar to evolution." Dawkins also writes that "Darwinian evolution, specifically natural selection . . . shatters the illusion of design within the domain of biology, and teaches us to be suspicious of any kind of design hypothesis in physics and cosmology as well." The question of "design" is semantically loaded, as it often in this context refers to the belief of people who reject evolution—i.e., "intelligent design." This is not the point. We know that evolution is how life-forms change through time. The question is whether evolution truly obviates a primal originator of the laws of nature—including the laws of evolution and the very important starting point of the evolutionary process. In fact, modern science has not been able to address these key issues.

Since Dawkins does not have advanced training in physics and mathematics, his arguments about the universe as a whole are easily disproved; in fact, no serious physicist would argue that a mechanism "similar to biological evolution" somehow operates in the purely physical universe. But we will more fully address Dawkins's scientific arguments against belief and expose their flaws later in this book.

Within a year after Dawkins's book appeared, the late Christopher Hitchens took up the New Atheists' aggressive charge, publishing *God Is Not Great* (2007). Hitchens seems to draw all his scientific points from Dawkins. A political and social commentator, not a scientist, Hitchens uses far fewer tools from science and more devices from journalism and philosophy in his censure of religion. His book is thus more similar to Harris's two earlier polemics, making less effective use of science than Dawkins, but with a wider-ranging condemnation of the evils of religion, blaming it for wars and genocides and tortures throughout human history. Like Dawkins, Hitchens also uses anecdotes from his own life to illustrate the overwhelming viciousness he sees in all religions:

> I was a member of the Greek Orthodox Church, I might add, for a reason that explains why very many people profess an outward allegiance. I joined it to please my Greek parents-in-law. The archbishop who received me into his communion on the same day that he officiated at my wedding, thereby trousering two fees instead of the usual one, later became an enthusiastic cheerleader for his fellow Orthodox Serbian mass murderers Radovan Karadzic and Ratko Mladic, who filled countless mass graves all over Bosnia.

**RECENTLY, SOME OF** the New Atheists have admitted that evolution and biological science alone are not enough to clinch their argument against faith, realizing that "creation" must also involve the inanimate world that precedes and supports life. They have therefore sought to extend their argument, as Dawkins started to do, to show that the universe itself could well have come about without a God. To accomplish this, they appeal to physics.

A prime example that characterizes this new trend in "scientific atheism" is the book *A Universe from Nothing: Why There Is Something Rather than Nothing* (2012), by Lawrence M. Krauss, which has become a best seller. According to the author, a physicist, his book was inspired by conversations with Hitchens when the latter had sought to understand physics and cosmology in order to try to use them more effectively in his writings. Alas, according to Krauss, Hitchens died before he could make use of such knowledge and before he could write the foreword he had promised Krauss. Dawkins, however, provided an afterword for the book.

Krauss's work argues that the universe came "out of nothing"—out of the sheer "laws of physics." He begins with a retelling of the widely known history of modern cosmology, from the discovery of the expansion of the universe by Edwin Hubble and his colleagues in 1929 to the equally important 1998 finding that the universe's expansion is accelerating (a discovery for which Saul Perlmutter, Adam Riess, and Brian Schmidt received the 2011 Nobel Prize in physics). Then Krauss employs quantum theory to conclude that the total energy in the universe is identically zero, and that hence the cosmos arose "out of nothingness." Just how he knows that the total energy is exactly zero when the universe is known to contain large amounts of "dark matter" and "dark energy," about which we know close to nothing, Krauss doesn't explain. He also doesn't reveal how the laws of physics themselves emerged. He says that "quantum mechanics tells us that there must be a universe." We will see, in fact, that quantum theory tells us no such thing.

Sam Harris published a further book in 2010, called *The Moral Landscape*, in which he argued that morality does not come from religion but rather evolves like other traits. This sentiment is also reflected in the writings of the philosopher Daniel Dennett, who has argued that evolutionary psychology explains the development of consciousness and morality and that there is no need for a "holy book" to tell us how to behave.

Atheism has been around at least since the ancient Greeks: followers of the Greek philosopher Epicurus (341–270 B.C.) believed that life came about by chance and that there is no supreme being that rules and cares for us. Like his contemporary Democritus, who proposed the first hypothesis that the universe is made of something like atoms, Epicurus too believed in the existence of such tiny basic elements of the cosmos, and he held that their random movements and interactions brought us the universe we see around us, with no design or creation or continuing control.

All societies have had believers and nonbelievers. Great thinkers such as Descartes were accused by their enemies of being atheists, even when they were not. And since society was often ruled by monarchs associated with and influenced by a church, atheists were often persecuted, heresy being a crime punishable by death. In the autos-da-fé in Spain, for example, Jews were burned at the stake as heretics. And the Italian philosopher, mathematician, and Dominican priest Giordano Bruno was burned to death in Rome in 1600 after the Inquisition found him guilty of heresy for believing that the sun was a star and that the universe contained other civilizations. But these persecutions of atheists, or people considered atheists even when they were not, took place centuries ago.

New Atheism is combative, aggressive, and belligerent against people of belief. Its proponents hold that religion is evil, and they state this belief loudly and clearly. Whether they are scientists or not, the New Atheists frequently employ science as their tool.

And these New Atheists—Dawkins, Krauss, the late Hitchens, Harris, and Dennett—are bound together under a powerful common purpose, and continually reinforce each other. The problem with the science in the books and lectures of the New Atheists is that it is not pure science—the objective pursuit of knowledge about the universe. Rather, it is "science with a purpose": the purpose of disproving the existence of God. The arguments these writers make, therefore, are often tendentious. They bend and distort science to further their own agendas in a way that is not too different from what a scientist in the pay of a pharmaceutical company might be doing in writing a favorable report about a questionable drug the company makes.

Top scientists, especially in the more mathematical fields

such as physics, tend not to get deeply involved in the atheist debates. They are generally more interested in pursuing science for the pure sake of learning about the universe, leaving philosophical debates about God to others (even though many of them may indeed be nonbelievers). Thus we do find books by major scientists who are sympathetic to atheism but are not part of the fervent atheists' cabal.

Among these writers is Stephen Hawking, whose book with Leonard Mlodinow, *The Grand Design* (2010), is a scientific odyssey to the frontiers of knowledge in physics and cosmology, supportive of the thesis that God is unnecessary but stopping short of the impassioned appeals to atheism of Dawkins, Krauss, Dennett, and Harris. I will address the arguments in the Hawking and Mlodinow book as well. Physical science, in fact, does not contradict the existence of God.

Hawking himself, one might add, has made use of "God," at least in the sense of the maker of the laws of physics, throughout his long career, with pronouncements such as "We are nearing God!" (on making a theoretical discovery in physics some years ago) and in the title of one of the books he has edited, *God Created the Integers* (2005). While certainly not implying any religious belief, Hawking's use of the word "God" is not something Dawkins, Harris, Dennett, or Krauss would ever do.

In a lecture given at Oxford University and reprinted in the British newspaper *The Guardian*, James Wood exposed some of the main flaws of the new "militant atheism," such as its relentless attacks on any kind of belief, its denying that there could ever be any value in personal religious practice, and its considering all Western religions abhorrent. Children, Wood remarked, are sometimes "stuck" in a strict literalism, out of which they eventually grow. However, he noted:

> The New Atheism is locked into a similar kind of literalism. It parasitically lives off its enemy. Just as evangelical Christianity is characterized by scriptural literalism and an uncomplicated belief in a "personal God," so the New Atheism often seems engaged only in doing battle with scriptural literalism. The God of the New Atheism and the God of religious fundamentalism turn out to be remarkably similar entities.

Scripture was never meant to be read literally. The original Hebrew text of the Old Testament and the Greek of the New Testament are elegantly poetic. But the aggressive New Atheists tend to take every word literally in ways that no rational reader would ever do. Christopher Hitchens, for example, quotes a passage from Joshua 10:12–13:

> Then Joshua spoke to the Lord on the day when the Lord delivered the Amorites before the children of Israel; and he said before Israel: "Sun, stand still upon Giv'on, and Moon in the valley of Ayalon." And the sun stood still, and the moon stayed, until the nation had avenged themselves of their enemies.

He uses this story to claim that "the script of the Old Testament is riddled with dreams and with astrology (the sun standing still so that Joshua can complete his massacre at a site that has never been located)." Jewish commentators of the Talmud, even as early as the Middle Ages, have read this passage allegorically, recognizing that for the sun and moon to actually *stop* their motions would require an extraordinary miracle, one that would fundamentally break the laws of physics, and that therefore this could not have happened. What is described here is a metaphor: the "stopped" sun and moon serve as a literary device to give the reader an idea about an exaggeratedly long battle.

How often do we hear someone describe a sublime moment, perhaps enjoyed on vacation on a beautiful beach on a tropical island, by saying, "Time stood still, we were having such a great time, sipping our drinks and watching an endless sunset"? No one would nitpick: "Was it really endless? Did time really stop?"

To be sure, the New Atheists were not the first to stick to uncompromising and unimaginative literalism. The Catholic Church, for one, historically adopted such literalism and unyieldingly defended it in the face of scientific discoveries and theories about the movement of the Earth, famously persecuting Galileo and many others for maintaining the nonbiblical heliocentric view. One would have hoped that the New Atheists, living in the twenty-first century, would know better.

In later chapters, I will counter the arguments of the New Atheists and show definitively that science hasn't disproved the existence of God. But first we should look at how religion came about in the first place, and see that in fact it was once connected with what may well have constituted very rudimentary "science"—inasmuch as it was an effort by early human societies to come to grips with nature and its laws.

# 1

# The Coevolution of Very Early Science and Religion

he earliest divinities emerged even before the dawn of civilization. These deities were abstractions of what very early humans had learned about the universe surrounding them in what may well be described as the first "scientific" deductions ever made. How do we know this?

Anatomically modern humans (that is, humans whose fossil remains are no different from the bones and skulls of present-day people) appeared on the continent of Europe sometime during successive waves of migration of hominids out of Africa, stretching several hundred thousand years into the past. Then, around 30,000 years ago—a date that roughly coincides with the disappearance of the Neanderthals (another human species, somewhat different from ours and now extinct), who had long lived on the Ice Age plains and hills of the continent—we sud-

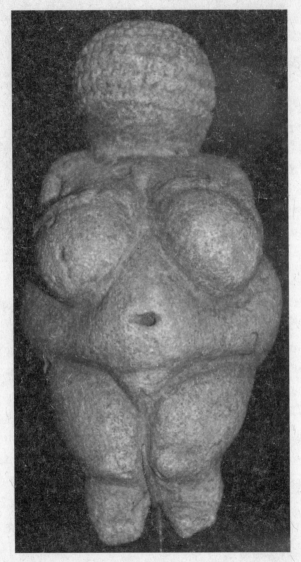

*The Venus of Willendorf, the celebrated early "Mother Goddess"*
*statuette found at a German excavation site near Willendorf.*

denly find the appearance all over Europe of curious-looking statuettes that have become known as "Venus figurines."

These are stone, wood, ivory, or clay images of nude women with large breasts and wide hips. They include the famous Venus of Willendorf, found in Austria; the Venus of Lespugue, from France; the Venus of Dolni Vestonice, from Moravia; and others.

And some Venus figurines may even be much older. The Venus of Tan-Tan, from Morocco, and the Venus of Berekhat Ram, from Israel—if proved man-made, rather than just stones that happen to closely resemble the human female form—would push back the origin of the goddess of fertility from 30,000 years ago to almost 250,000 years in our past.

Archaeologists believe that the Venus figurines are images of the first deities ever conceived. They are universally understood to represent *fertility*—something that was extremely important for the survival of the small groups of nomadic humans living as hunter-gatherers in Paleolithic Europe. More broadly, in these fertility gods, we see that the human impulse to worship the creative/generative force of the universe extends to the earliest days of our species. One is tempted to say that spirituality is in our nature.

In addition to "portable art," which includes the Venus figurines made by these early people, the Upper Paleolithic (the Late Stone Age, 40,000 to 10,000 years ago) is also characterized by magnificent cave art that can be seen in hundreds of deep caverns in France, Spain, and (to a lesser extent) Italy. These caves include the famous Lascaux, in the French region of Dordogne;

Pech Merle, in south-central France; Niaux, in the French Pyrenees; Altamira, in western Spain; and many others.

The art found in these caves consists mainly of images of animals, usually painted but sometimes engraved on the walls. The images are typically found in the deepest, most inaccessible parts of the caves. Often one would have to walk or crawl for hundreds of feet (in the cave of Niaux, for example, this entails an arduous walk of half a mile through often very slippery and narrow stone fissures deep inside the mountain) to reach the deep "halls" in which the paintings are found. There, one sees beautiful images of bison, horses, ibex, deer, mammoths, and wild cattle (called aurochs).

In the cave of Chauvet in the French Ardèche, dated to at least 35,000 years before the present, is found, among scores of stunningly beautiful images of horses and bulls and ibex, a singular charcoal image of a bull superimposed on the naked lower half of a woman with an exaggerated pubic triangle.

Paleoanthropologists have identified this image as a fertility symbol similar to the Venus figurines. The bull in ancient art often represents a force of nature, and the juxtaposition of the naked lower part of a woman's body with the bull on top may represent the union of two deities: a powerful god that stands for the forces of nature such as wind and fire, earthquakes and storms, and the female goddess of fertility.

Experts theorize that early humans, still living on the frozen prairies as hunter-gatherers, closely observed nature and deduced the connection between sex and fertility. To promote fertility, they idolized the female form and the female genitalia. Images of vulvas dot many of the caves of Europe decorated by Paleolithic humans, dating from about 40,000 to 11,000 years ago.

The French anthropologist André Leroi-Gourhan has proposed that all these images are a manifestation of perhaps the earliest religion in the human past. His theory takes into account finds from the French caves of Gargas, with Paleolithic handprints; Niaux, with its rich animal symbolism; Lascaux, which Leroi-Gourhan dubbed a veritable "cathedral of prehistoric religion," abounding with images such as a leaping cow and wild horses; and the cave of Altamira in Spain—the earliest to be discovered, in the late nineteenth century—with a ceiling covered with images of bison apparently in rut, which may also symbolize fertility. Many of the prehistoric caves in France also bear strange signs that perhaps indicate the use of language, although they remain undeciphered. The Neolithic Revolution, which took place around 11,500 years ago, catapulted this very early kind of science-religion mix to the fore. The French archaeologist Jacques Cauvin describes in his groundbreaking book *Naissance des divinités, naissance de l'agriculture* ("The Birth of Divinities, The Birth of Agriculture," 1997) how the advent of agriculture, with the first cultivated crops and the earliest domestication of animals, which took place in the Middle East at the end of the Paleolithic, inspired new kinds of religious observance.

The rise of agriculture on Earth is a true marvel—and it

happened in at least two separate locations in the world: the Jordan Valley in the Middle East and the coast of Ecuador in the Americas. Around 11,500 years ago, Natufian people—named after Wadi El Natuf, where archaeological remains of their culture have been discovered in caves above the Jordan Valley— learned how to harvest wheat and barley that happened to have undergone a rare mutation: the grains did not fall randomly to the ground after maturing, but somehow remained on the stalk, as if waiting for the reaper. We don't know how the Natufian people were able to find grains that had this rare mutation, exactly what was needed for agriculture to succeed. Perhaps they experimented for many years to learn how to cultivate the mutated forms.

The invention of agriculture and the domestication of animals such as goats, sheep, and cows, which happened at the same time in the Middle East, brought new meaning to the older Paleolithic concept of fertility, which until then had been limited to human populations and those of hunted animals. Natufian and later agricultural communities in the region of modern-day Israel, Jordan, Syria, and Turkey became acutely aware of their dependence for survival on the fertility of the land, the continuing reproductive ability of domesticated animals, and the continuity and growth of the human population.

According to Cauvin, this is why we see a veritable explosion of female figurines in the entire newly agricultural communities of the Middle East and Anatolia eleven to nine millennia before our time. In contrast with the hunter-gatherer nomadic cultures of Europe of the Upper Paleolithic, which produced the magnificent cave art and left us the older Venus figurines, Neolithic female images are more abstract and stylized. According to Cauvin, what we have here is the emergence of a new religious symbolism.

At the sites of ancient settlements such as Netiv Hagdud in Israel, Mureybet in Syria, and Salibiya in Jordan, for example, archaeologists have discovered hundreds of stylized stone female figurines in which generally only the eyes, the breasts, and the pubic triangle are evident. They believe these to be images of a new goddess of fertility responsible for a bountiful harvest, abundant human births to keep the population stable, and animals to provide meat, milk, and wool. Coincidentally, images of bulls abound during this era as well, which archaeologists believe symbolize the powers of nature.

What we see here is people *learning from nature,* which is something that can well be identified as a rudimentary kind of science. These people learned that the idea of fertility permeates nature as a whole, and they chose to symbolize the general concept of fertility—so essential to their survival—in the form of a woman's reproductive and infant-nurturing organs.

In *After the Ice: A Global Human History, 20,000–5000 BC,* Steven Mithen describes some of the most exciting evidence of emerging prehistoric religion, found inside ruins of stone houses at the site of prehistoric Jericho:

In the third room three clay figurines, each female and about 5 centimeters high, stand against a wall. One is particularly striking—it is dressed in what seems to be a flowing robe that has been sculpted with arms bent, so that one hand is placed below each breast. Next to them is what looks like a human head . . . or at least a skull on which the face has been delicately modeled with plaster.

Çatalhöyük, in southeastern Turkey, is an unusually rich archaeological site dating from 9000 B.P. (before present). Here were found square houses packed together, and in about a third of them were found wall decorations and other art interpreted by archaeologists as religious: paintings of leopards and bulls, and Venus figurines; these buildings have been conjectured widely by scientists to have been shrines. But the most impressive and enigmatic image found here was an eight-inch-tall stone carving of a woman with heavy exposed breasts seated on a throne flanked by two lions, her hands on the tops of their heads. Archaeologists have dubbed her "The Mother Goddess" and she may represent a fertility goddess dominating the other forces of nature, represented by the lions.

From about the same period as Çatalhöyük is a settlement called Nea Nikomedeia in Greece. Here archaeologists discovered an ancient shrine with several Venus figurines made of terra cotta, which seem a clear adaptation of the earlier figurines

*The enigmatic Mother Goddess, flanked by two animal images, from the Neolithic site of Çatalhöyük in Turkey.*

found in the Middle East. These Venuses have more stylized bodies with narrow waists and prominent buttocks.

At Tel El Ubaid in southern Mesopotamia (present-day Iraq), there flourished at around 7900 B.P. a culture that archaeologists named the Ubaid, which within five hundred years spread to the whole of Mesopotamia, replacing an earlier culture called the Halaf. The Ubaid people built many mud-brick temples to their gods, each with an altar and an offering table.

In the summer of 2008, I accompanied the renowned Harvard archaeologist Ofer Bar-Yosef, who had unearthed many fossils of early humans and Neanderthals in the caves of Carmel in Israel, to a newly excavated Neolithic site in the Galilee called Yiftahel. Yiftahel dates from around 9000 B.P. and has ruins of square houses similar to those found at Çatalhöyük in Turkey. Bar-Yosef wanted to visit this site because of a surprising discovery that had just been made here. We turned off the main highway from Haifa and onto a dirt road, and after a few minutes were greeted by the archaeologist in charge of the site.

Everyone was excited to show us the new finds. We carefully made our way on wooden planks placed high above the now-exposed remains of the walls of prehistoric houses constructed of large stones piled one on top of the other in rectangular arrays. At the bottom of one of the houses, several people were huddled over cardboard boxes filled with fluffy protective packing material. They were carefully boxing two exquisitely preserved skulls, each plastered over with masklike material still showing the remnants of ancient paint.

These painted skulls, just about ready to be shipped to a lab for analysis, had been found beneath the floors of two of the houses. Both were clearly plastered over in order to form an image of a face. Similar artifacts had been discovered elsewhere, including in Jericho, as mentioned by Mithen, and attracted great attention. This was evidence of what archaeologists believe was an ancestor cult.

These ancient peoples are believed to have removed the skulls from the skeletons of their parents or grandparents, plastered them in wet clay, and painted the faces of the deceased persons on them. Archaeological evidence suggests that they kept these plastered skulls under the floors of their homes as idols or deities, perhaps to watch over them and bring them good fortune.

What was so exciting was that none of the other sites in this particular area had any evidence of the ancestor cult—only fertility images and bulls. But in Jericho; at the Nahal Hemar cave in the Judean Desert; in the Hula Valley, in the north of Israel; and in Çatalhöyük, these bizarre plastered skulls had been discovered. The work of covering the skull with plaster, filling in the nose and eyes, adding lips, and painting it turned a skull into something eerily resembling a human head.

This practice is viewed as a forerunner of the Christian idea of rising from the dead, or the notion of an afterlife present in many modern-day religions. It was likely derived from fear of death and the incomprehensibleness of dying—something that affects all of us humans and is a clear motivator for religious belief.

Just before the beginning of the Chalcolithic period (6500–5500 B.P.)—marking the first use of metal, specifically copper, in addition to stone tools—we see a sharp increase in the creation of *places* dedicated to the gods. These sanctuaries are locations that are marked off with stone walls and are divided into outer and inner areas. The innermost area is the "holy of holies" of the sanctuary, the place where the divinity is supposed to reside. The sanctuaries are the forerunners of the later temples, synagogues, churches, cathedrals, and mosques of modern-day religions.

The figurines representing fertility and statuettes of bulls and rams representing forces of nature were now placed in the sanctuaries people made for them. One of the oldest sanctuaries ever found is the Leopard Temple, built around 7500 B.P. in the Uvda Valley, in the Arava wilderness of Israel. This is a rudimentary temple built of stacked stones, in front of which is a large array fifty feet long by fifteen feet wide of statues depicting leopards hunting an oryx. Some archaeologists hypothesize that the leopards represent divinities residing in this temple-sanctuary.

Chalcolithic farmers and copper miners are believed by some experts to have performed ceremonies honoring their gods as well as celestial bodies such as the stars, sun, and moon. During this period we see the emergence of a class of priests or shamans responsible for leading religious ceremonies. There is a proliferation of ivory and wood Venus figurines during this time, but also new ornamental copper objects that were used in worship and ritual. At the Cave of the Treasure at Nahal Mishmar in the Judean Desert was found a trove of copper and elephant ivory objects used in religious rituals: mace heads, scepters, and crown-shaped objects. A total of 429 such items, wrapped in a woven mat, were discovered here in 1961, dating from about 6000 B.P.

Most of the clay figurines from the Chalcolithic are female fertility goddesses. Some are clay or stone images that show only breasts and hips. The male god of this period is represented mostly by images of a ram. Libations (liquid offerings to the gods) began at this time, with the production of a new object characteristic of the Chalcolithic: a cornet, or cone-shaped vessel. At another Chalcolithic sanctuary, at Gilat in the Negev Desert, were found figurative libation vessels; in one carving a ram carries cornets.

Ancestor cults spread during this period, and more plastered skulls have been discovered at other Chalcolithic sites. Burial became highly ritualized, and is characterized by the making of clay ossuaries: elaborately decorated clay containers for bones. Burial sites were now separate from dwellings and found outside the cities. Ossuaries were placed in caves dug especially for this purpose. At a Chalcolithic site in Peqi'in, in the Upper Galilee, hundreds of elaborate clay ossuaries, finely decorated with images of animals and imaginary beings, have been discovered.

People of this early era in human development clearly had a belief in the afterlife and devoted great resources to the cult of the dead; the nature of the artifacts they left—which connected the dead with life and with daily activity—make it evident that they thought the dead had powers to protect the living. Devotion to one's ancestors became a means of guaranteeing one's protection and patronage.

In the early Bronze Age, 5500–4300 B.P. (3500–2300 B.C.), figurines took many forms. Religion was an important part of everyday life during this period. Religious officials in the societies of the Near East occupied a high social status, and sacred rites were held regularly in temples and palaces, both of which were always built at the city center.

The Egyptians and Babylonians created pantheons with gods and goddesses representing various powers of nature, and we see similar pantheons later appear in Greece and Rome. In the Greco-Roman pantheon we recognize a continuation from the female fertility goddesses of the Neolithic in the form of the goddess of love, Aphrodite (in Greece), later known as Venus (in Rome), as well as other deities who enjoy intricate relationships with one another and with humans. The bull continued to serve as an image of divine power, as we see in the Minoan myth of the Minotaur.

The physicist Max Jammer has traced the idea of physical force seen in nature as it found its way into religion. In *Concepts of Force* (1999), he writes:

> With the progressive organization of early society
> into urban civilization the concept of a capricious

interplay of forces behind the ever-shifting phantas-
magoria changed into the idea of a systematized hi-
erarchy of forces of nature; eventually "force" as such
was personified into a spirit or a god of overwhelming power. Such personification was characteristic of
ancient mythology which, as the only body of systematized thought of those times, was not only the
cosmology but also the "physics" of the prescientific
stage.

Jammer describes the concept of a god as an embodiment of
a physical force in the mythology of ancient Egypt, as evident
in Papyrus Harris I, an Egyptian manuscript kept at the British
Museum and dated to 1185–1153 B.C. Jammer writes that "nht"
in this papyrus is a divinity that embodies the abstract idea
of physical force. Forces of nature, "seen as personal beings,"
emanate from the "nht" deity and carry out the action of the
force.

It is striking to note a similarity between this ancient idea
of a physical force and modern physics. In particle physics, a
force, such as the weak nuclear force acting inside the nucleus
of an atom, is effected through the action of a particle—called a
boson—that makes that force active. Just as "nht" acts by send-
ing "personal beings," the weak nuclear force that acts inside the
nucleus "sends" a W or a Z boson to carry out its action within
the nucleus.

Ancient peoples saw in the stars a cosmic order, and hence

in their mythologies some of the early deities come from the heavens. Images of stars played a key role as divine representations from early antiquity to the establishment of the Western religions: In ancient Mesopotamia, the four-pointed star represented the powerful Sun God. In Canaan, the morning star represented one of the children of the gods, and the Hebrews of the reign of the Maccabean king Alexander Jannaeus (103–76 B.C.) minted coins with a five-pointed star, representing divine protection. Early Christians adopted the image of a six-pointed star, representing the joining together of Jesus's Greek initials, I and X. It was this image that later became the Star of David. And the Star of Bethlehem, perhaps Venus, plays an important role in the Christian narrative as having led the magi to the newborn Jesus.

In ancient Egypt, as later throughout the world, religion was established as the guardian of morality. For instance, in the ancient Egyptian tale of Setne, the sun god, Ra, sends an emissary to carry out his force to punish a robber. A force responsible for physical action thus assumes a moral role.

The Canaanites had a powerful storm god, and the earliest sanctuaries in the Canaanite realm (today's Israel) were dedicated to him. In ancient Mesopotamia, we also find a storm god, called Enil, who is characterized by inconsistency, vicissitudes, and dynamic action resembling those of an unpredictable force of nature. In contrast, Anu, the sky god, is placid and serene and represents the concepts of immutability and permanence. Again, ancient divinities were abstractions of forces seen in nature, mixed with human characteristics and an overlay of an emerging morality—punishments are meted out by these forces for "sins" such as theft and murder. It's clear that morality and a code of behavior in a civil society originate with very early religions. Early science—i.e., an understanding of nature and its forces—also goes hand in hand with the development of spiritual practice and moral codes.

Jammer traces the genesis of the God of the Old Testament as a consolidation of gods representing different physical forces into a single deity. He shows that one of the Hebrew names for God, Shaddai, originates in the Semitic root *shadda*, which means "to have great force." Similarly, an early Arab deity is called *Al-Uzza,* which means "the mightiest," and Jammer claims that the modern Muslim name for God, Allah, is a derivation of this ancient title.

But in their drive to establish monotheism as a universal form of belief, the evolving Western religions lost some of their rootedness in the natural world. While originally religion constituted a deification of forces seen directly in nature—fertility and physical phenomena—with the advent of Judaism and the continuation to Christianity and later Islam, we see more "revelations," miracles, and supernatural beings such as angels. Religion progresses to the wrath of the Hebrew God, the torture and death of Christ, and Muhammad's conquests.

As Christianity spread in the first few centuries after Jesus and the destruction of Jerusalem by the Romans, small Christian communities multiplied throughout the Roman Empire, along with Jewish ones and a variety of pagan sects. In an edict of A.D. 380, the emperor Theodosius established Christianity as the official religion of the empire. The modest-sized early Christian communities had an initiation ceremony, baptism, called "an enlightenment" or "a rebirth"; their members pledged an allegiance to Jesus Christ; and they took part in a communal meal called the Eucharist.

These early Christian communities were drawn together by the hope of salvation and the belief in a single benevolent God who controlled all things, from the creation of the universe to the daily lives of men. Common rituals gave a purpose and a meaning to life. Religion in this form far transcended the natural world people saw around them, and answered to the ongoing human quest for meaning and understanding. The Christian God, like the earlier Jewish God, represented a unified, single force governing everything, including morality. In this way, an understanding took hold that humanity, the Earth, and the cosmos are one interconnected system.

Early Christians avoided creating images, still following the old Jewish prohibition on making art of any kind. But in the third century, Christianity moved away from such connections to Judaism, and Christian art emerged resplendent. During this period, mosaics depicting biblical scenes multiplied all over the Mediterranean basin, from Palestine to Egypt, Greece, and

Italy. The purpose of the new art was to spread the Gospel: to literate people, this was done through reading, and to the illiterate through pictures. The early mosaics we enjoy in museums and at archaeological sites today owe their existence to emergent Christianity, which continued to provide the subject matter for most of the great Western art through the Renaissance. Even when taking into account the fact that the church was the major benefactor of artists for centuries and thus shaped the content of art, it's undeniable that spiritual inspiration has fueled creativity, from Michelangelo to Mark Rothko.

The relation between religion and science also persisted. At least some early Christian thinkers understood important facts about the physical universe. In their book *The Grand Design,* Stephen Hawking and Leonard Mlodinow discuss the writings of the great Christian thinker and philosopher St. Augustine of Hippo (A.D. 354–430), who lived during the consolidation of Christian identity throughout the Roman Empire. Augustine worried about the question of what God was doing before he created the universe. His answer was surprisingly close to that provided by modern physics. According to Augustine, time itself was created when God made the universe, so the question is moot: before the creation of the cosmos there was simply no time. This agrees with what cosmologists believe today, that is, that both space and time were created in the Big Bang and that therefore there was no "time" earlier than the Big Bang.

It's widely understood that the figure of the Virgin Mary

is related to the fertility goddesses of mythology and prehistory. For example, the Collyridian sect, which flourished in the fourth century, revered Mary as the mother goddess of antiquity. The ancient Egyptian goddesses Hathor and Isis are a manifestation of the fertility goddess or mother goddess, and a worship of Isis persisted well into the first few centuries A.D. Isis worshippers are believed to have reconciled themselves to Christianity by adopting the Virgin Mary as a mother goddess. Fertility or mother goddesses were part of mythologies in the Americas as well. The Aztec had Toci, the mother of the gods; and after the Conquistadors introduced Christianity, in addition to the Virgin Mary many Native American cultures worshipped a mother goddess called Pachamama.

The saints of the Catholic Church resemble the Greek and Roman pantheons—each saint with powers and a specialty similar to a god. But overall, Westerners chose monotheism. The Hindu religion of India, by contrast, had its own pantheon in which various deities influenced the processes of the natural world, including sex and fertility. Science, including mathematics, is entwined throughout this mythology. This connection is especially evident in the mysterious temples of Khajuraho.

In 1838, a captain in the British military named T. S. Burt was exploring the jungles of the Indian state of Madhya Pradesh, some four hundred miles southeast of Delhi, with his company of Bengal Engineers, when he and his men came upon a group of ancient temples that had been taken over by the jungle at a location called Khajuraho. What they saw here stunned Burt. He noted that these temples were among the finest he had ever seen, but he was at a loss as to how to describe in his logbook the graphically erotic art at Khajuraho. About a tenth of all the magnificent ancient statuary at Khajuraho depicts sexual situations. In the West, we don't see graphic sexual imagery in public places—and certainly not in traditional places of worship. But the statues Burt saw, of men and women engaged in all kinds of sexual activity, were right on the outside walls as well as in the chambers of what were clearly places of religious worship: Hindu and Jain temples built a thousand years ago.

Burt wrote in his diary: "I found seven Hindoo temples [there had once been eighty-five of them, of which twenty survive, both Hindu and Jain] most beautifully and exquisitely carved as to workmanship, but the sculptor had at times allowed his subject to grow a little warmer than there was absolute necessity for his doing." All temples in this area, "a stone's throw away from each other," as Burt wrote, have statues adorning their walls. These statues depict everyday life, as well as images of deities. But the erotic scenes strongly dominate these temples because they are so explicit, and so unexpected from a Western perspective.

Equally inexplicable was the existence of a piece of complex mathematics. On the doorway of the Parsvanatha temple, a Jain temple in the eastern group of temples at Khajuraho, is an inscribed magic square, with Hindu numerals (some of which are like our own and some quite different). This is a four-by-four square with the following digits (here transcribed using modern numerals):

| | | | |
|---|---|---|---|
| 7 | 12 | 1 | 14 |
| 2 | 13 | 8 | 11 |
| 16 | 3 | 10 | 5 |
| 9 | 6 | 15 | 4 |

Now notice some amazing facts: The sum of every horizontal row is 34; so is the sum of every vertical column; the sums of the two diagonals; and so are the sums of all two-by-two squares at the four corners of the larger square; and also the sum of the central two-by-two square. The temple bearing this curious inscription is definitively dated, by an inscription, to A.D. 954. So as early as the mid-tenth century, the people who built and worshipped here understood how to construct such sophisticated pieces of mathematics. The Khajuraho magic square is one of the oldest known four-by-four squares, though there are earlier three-by-three squares from China and Persia.

This magic square and some of the less-explicit images from the outside of the Parsvanatha temple are shown in the photographs below. The Japanese mathematician Takao Hayashi photographed this square in the 1980s but had lost information on its exact location; I rediscovered it after a long search in 2011.

So why did tenth-century Indians place a magic square on the doorway of their temple? The answer is no more within our

*The magic square and the decorated front face of a*
*tenth-century temple at Khajuraho, India.*

grasp than is the purpose of its plethora of highly erotic art. Perhaps the priests believed that numbers held great powers that they hoped to harness. Sex and mathematics may have been viewed as manifestations of the forces of nature these people wanted to worship. The earliest zero in history was found in an inscription from the mid-seventh century discovered in the Sambor temple in Cambodia, again showcasing the connection between mathematics and worship in Eastern religions.

The great temple of Angkor Wat, in northwestern Cambodia, which was a center of both Hindu and Buddhist worship, contains many images of *apsaras*—alluring female deities often identified as fertility goddesses.

*The alluring* apsaras—*carvings of goddesses decorating the facade of the eleventh-century temple of Angkor Wat, Cambodia.*

Angkor Wat is the largest temple ever built, with an area of 203 acres. It was constructed in the twelfth century, at the same time as another great religious building: the cathedral of Notre Dame in Paris. The twelfth century was thus the time of intensive construction of great religious edifices in the East as well as in the West. In Europe, in particular, the great cathedrals are grand displays of religious faith that remain with us today.

In his book *The Light in the Church: Cathedrals as Solar Observatories,* J. L. Heilbron explores one relationship between religion and science, showing that the way light enters some of the great Gothic cathedrals indicates that they were originally used as solar observatories to determine the exact date of Easter. The Catholic Church, Heilbron argues, encouraged advances in science—even in astronomy, despite the later prosecution of Galileo. Heilbron shows how science was used in the building of the great cathedrals, as well. Flying buttresses, for example, are an immense technological development of Gothic architecture in the twelfth century, which enabled the construction of cathedrals that were much larger, supported heavier roofs, and had more window space to allow light to enter. Flying buttresses are arches that lead away from the load-bearing walls to the ground, providing much more support than the less-efficient buttresses in earlier cathedrals, which were attached to the walls. The new buttresses enabled the construction of the magnificent Gothic cathedrals of Notre Dame in Paris, Chartres, Reims, and other great European cathedrals we admire today.

The Crusades, taking place around that time, with their

bloodshed and mayhem, are an example of the evil done in the name of religion. But morality, too, developed in the Middle Ages. The Talmud, for example, though written in the third century, was expanded throughout the Middle Ages. It provides a moral guide to law and life, as did the books written by the medieval Jewish doctor and philosopher Moses Maimonides, whose writings on morality are still followed by many religious Jews today, and who has also studied science. Already in the twelfth century, he explained that the literalist approach to Scripture was wrong. Eight and a half centuries later, the New Atheists still go back to the literalist image of the Old Testament God as an old man with a long white beard and a bad temper instead of considering what Maimonides proposed, that God is the immensity of power in nature: the force behind creation, mathematics, art, humanity, and decency.

And already in the second century A.D., the Talmudic sage Hillel the Elder famously summarized the message of the Old Testament in one sentence: "What is hateful to you, do not do to your fellow man; this is the entire Torah, the rest is commentary." The message of love for our fellow humans was thus propagated through the ages as Judeo-Christian morality, elucidated by Jewish and Christian thinkers to our time. The view that morality does not relate to religion—a tenet of Dawkins (should we "cede to *religion* the right to tell us what is good and what is bad?"), Harris, Hitchens, Dennett, and other New Atheists—is simply false.

**BECAUSE THE CATHOLIC** Church sought to preserve a literal interpretation of Scripture and promote the central role played by Earth within the wider universe, it chose to follow the ideas of a nonscientific, pre-Christian philosopher: Aristotle (fourth century B.C.), whose philosophy relied on reason more than experimentation, resulting in errors such as his belief that heavy objects fall faster than light ones (disproved by Galileo centuries later). Aristotle's geocentrism, and his insistence on the immutability and constancy of nature, became pillars of Catholic belief that would stand in the face of scientific developments from the late sixteenth through the seventeenth century, culminating with the trial of Galileo. But science was undeterred, and the important scientific work of Galileo was continued by other great thinkers: Kepler, Descartes, Leibniz, and Newton.

Before we go on to look at the important achievements of these great minds, who made physical science—the means by which we understand nature and physical phenomena—what it is today, we need to give a closer look to the archaeological evidence for the events described in the Bible. In this chapter, we surveyed the development of religious ideas as having co-evolved with basic scientific notions and then diverged from science as Western religions matured. A key element of the New Atheists' attacks on religion has been the casting of doubt about the entire biblical narrative. In the next chapter, therefore, we examine the archaeological and historical evidence for the foundations of Western religion.

# 2

## Why Archaeology Does Not Disprove the Bible

Richard Dawkins claims in *The God Delusion* that there is no shred of evidence for any of the stories of the Bible. Christopher Hitchens concurs with this view in *God Is Not Great*, claiming that modern archaeology has disproved biblical history ("none of the religious myths has any truth to it, or in it"). In fact, this is not quite true. Biblical archaeology is a thriving field, which has brought us troves of evidence for ancient settlements in the Holy Land and for some of the scriptural events (although nothing supernatural) that took place there.

The biblical Abraham was born in the city of Ur in Mesopotamia. By tradition he smashed the baked clay figurines representing Babylonian gods and goddesses and professed his new belief in one God, engendering the rage of the heathens, which made it necessary for him to flee, eventually to the land of Canaan.

The Bible tells us the rest of the story of Abraham's travels, his marriage, the birth of his son Isaac, followed by Jacob, the immigration of their descendants to Egypt, the Exodus, the conquest of Israel, the dynasties of the Hebrew kings, the life and teachings of Jesus, and the Crucifixion.

But what is the evidence for the biblical stories?

Inscriptions relating to the kings of Judea and Israel have been found at sites throughout Palestine—for example, many impressions of Hezekiah's "LMLK" ("to the King") seal, dating from the eighth century B.C.—and can now be seen in several museums and publications.

There are also remains of many ancient sites described in the Bible, including the Temple of Solomon: In February 2010 a team headed by Hebrew University archaeologist Eilat Mazar uncovered a seventy-meter-long and six-meter-high wall, dating from the tenth century B.C., lying directly beneath the first-century A.D. Wailing Wall in Jerusalem. This is exactly the location and estimated date of Solomon's First Temple.

One exceptional find from the Second Temple in Jerusalem is a four-foot-long marble lintel bearing the inscription "To the trumpeting house." It once stood in the Temple of Herod (the first century B.C. Second Temple, destroyed by the Romans in A.D. 70), and its purpose was to direct the priests to the location from which a trumpet was to be blown in order to signal the beginning and the ending of the Sabbath. So we know definitively from this find that the Second Temple existed in Jerusalem, and that people worshipped there.

In addition, a major component of Jerusalem's complicated ancient water system (called Warren's Shaft after the Englishman Sir Charles Warren, who discovered it in 1867), dating from as early as the eighteenth century B.C., when the city was Jebusite, can still be visited in Jerusalem today. It confirms the biblical description of King David's conquest of the city shortly before 1000 B.C., when David's men sneaked into the water tunnel at its source at the Gihon spring and entered the well-fortified city through it (as described in 2 Samuel 5:8).

The ruins of Jericho, excavated by the British archaeologist Kathleen Kenyon in the 1950s, provide us with support for the scriptural story of the city's destruction: the level of Jericho's remains that corresponds to the era when biblical scholars believe Joshua was conquering the land of Canaan, around 1200 B.C., has been shown to contain large swaths of burnt buildings. So while the literal interpretation of Scripture—that the sound of Joshua's trumpets made the walls of Jericho come tumbling down—is certainly doubtful from our present understanding of the laws of physics, we do have good archaeological evidence that this ancient city, one of the oldest in the world (its foundations, excavated by Kenyon, go back to around 9000 B.C.), was indeed destroyed in biblical times.

On May 24, 2012, a find of great importance was announced in Israel. The team led by archaeologist Eli Shukrun, who had been excavating around the ancient wall of the city of Jerusalem, had discovered a one-inch seal bearing the words "Beit Lechem"—Bethlehem—dating from the eighth century B.C. The full text of the seal reads: "On the seventh, Beit Lechem, to the King." This was a tax writ sent to the king of Judea (Hezekiah, Menashe, or Joshiahu) from Bethlehem to Jerusalem. The seal proves the existence of the town of Bethlehem as early as the eighth century B.C.

Early Judaism was not fully monotheistic, and various mosaics have been discovered in northern Israel showing a Jewish menorah accompanied by Greek mythological figures. These artifacts date from the first and second centuries B.C. Earlier sanctuaries to various gods have also been found in Israel. The change to monotheism was gradual.

The first sanctuaries—places built to house the gods, as mentioned earlier—appear in the land of Israel around 5500 B.C. These sanctuaries provide key archaeological evidence for our understanding of how religion developed in ancient Israel and also confirm the biblical accounts of the conflicts that raged among the nations of the region over religion and land: Hebrews, Canaanites, Philistines, Jebusites, Amonites, Aramites, and others fought against one another and destroyed each other's sanctuaries.

In early Iron Age Palestine (1200–1000 B.C.), the archaeological record reveals a plethora of religious practices, marked by symbols of power: lions, bulls, and the "Tree of Life." These are images that stand for protection and fertility. A unique artifact discovered at a place called Taanach bears a striking resemblance to the accompanied mother goddess of Çatalhöyük created several millennia earlier. The Taanach find is a pottery stand with many motifs including a nude woman flanked by lions and a calf, believed to symbolize yet another deity. On top of the display is a winged sun disk, perhaps representing the sun god.

Many of the events of the Bible and the reigns of the Israelite kings have been dated with extreme accuracy. This has been possible because of a concurrence of datable astronomical events recorded in Mesopotamia with stories told in the Bible. Assyrian and Babylonian astronomers kept meticulous records of total solar eclipses. In particular, an eclipse was recorded to have taken place during the Assyrian month of Simanu, in the year of Bur-Sagale (years were given names of governors and other high officials), which modern astronomers have been able to fix as June 15, 763 B.C. This allowed historians to establish large portions of the Assyrian chronology.

Now, a large commemorative inscription, definitively dated using the solar eclipse date to 701 B.C., recounts the story of Sennacherib's conquest of Jerusalem that year. The inscription describes it as follows: "Sennacherib went against the Hittite-land and shut up Hezekiah the Jew . . . like a caged bird in Jerusalem, his royal city." In 2 Kings 18:13, this same event is described from the Hebrews' point of view as taking place in the fourteenth year of King Hezekiah. This solid correlation between the Bible and an Assyrian source has enabled us to determine the entire chronology of the Hebrew kings described in the Old Testament.

We thus know that the founder of the Hebrew dynasty, King David, reigned around 1000 B.C. Other confirmations of parts of the story come from archaeological finds, among them an inscription discovered at Tel Dan in the north of Israel that describes the "house of David."

At the time of the establishment of the Israelite kingdom, the Hebrews lived as seminomadic groups in the hill country of Judea and Samaria, where they later started building more permanent settlements. Archaeologists have uncovered several such villages, each consisting of a ring of houses surrounding a common area used for livestock. Religious practice took place on *bamot*—stages or high places—a number of which have been discovered by archaeologists. In one of them a bronze statuette of a bull was uncovered, reminiscent of the Golden Calf of Exodus as well as golden calves that the Bible says were placed by King Jeroboam I at Dan and Bethel. This practice seems to be an intermediate step between idolatry and monotheism, and it is characteristic of Israelite finds from about 1000 B.C.

After the reigns of David and Solomon, which commenced around 1000 B.C. and ended half a century later, the Israelite kingdom split in two. Judah was the southern kingdom, which continued to be ruled by the house of David, and Israel was the northern kingdom, situated in the fertile hills of Samaria. Unique archaeological evidence for the house of David has come to us in the form of an Aramaic inscription (Aramaic being the lingua franca of the Near East of this time), which was part of a monumental stone slab commemorating the military victories of Hazael, the king of the rival nation of Aram. This piece of ancient writing on a stone is a reference to the dynasty of David and to the biblical kingdom of Judah. The inscription is King Hazael's boast of killing two kings, Joram of Israel and "Ahaziah of the House of David." While Hazael's name does not appear in the Bible, the fact that an archaeological find confirms the existence of the Israelite dynasty of David is of immense importance in confirming some accounts of the Old Testament.

Another important piece of archaeological evidence for the historical kingdom of Judah is an epitaph marking the burial place of King Uzziah, who ruled in the eighth century B.C. Since this king was a leper, he could not be buried in the royal tombs in Jerusalem, which is why this grave was found outside the city walls.

Further evidence comes from the gate of the fortress of Hazor, in the north of Israel, built by King Ahab of Israel. Only the right half of this large gate has been unearthed by archaeologists. It consists of two columns with carved capitals that support a massive lintel. The two capitals are decorated with a palm-tree motif that has been identified as a fertility symbol, common in Near Eastern art of this time.

In the Israelite capital of Samaria, archaeologists have discovered many decorated ivory items that belonged to the royal family. These objects are mentioned in the Bible: the prophet Amos complained about the "ivory beds" that epitomized in his mind the "depravity and corruption" of the royal family and the aristocracy of the Israelite kingdom.

In the City of David in Jerusalem, about fifty seal impressions—similar to the signatures of our time—have been discovered from the time of the kingdom of Judah. These are believed to constitute the remains of an administrative archive belonging to the royal family. Similar seals connected to the commander of the Judahite fortress of Arad, in the south of the kingdom of Judah, have also been found. The letters themselves were written in biblical Hebrew on potsherds found in the area. A seal made of agate, belonging to a high official, "Yaazanyahu, servant of the King," was discovered at Tell en-Nasbeh, dating to the late eighth century B.C.

The First Temple—the Temple of Solomon—lies under the ruins of the Second Temple, the one rebuilt by Herod the Great centuries later. We have little archaeological evidence for the First Temple itself, aside from the wall discovered in 2010, since digging in this area would cause great controversy—leaders from several faiths are opposed to such digging because they believe that the holy site should not be disturbed. However, many items used in worship in the First Temple have been discovered and dated to this period. These include incense shovels, inscribed offering bowls, and pottery shards bearing the names of priestly families. An ornate bronze ritual stand decorated with lions, oxen, and cherubim, reflecting descriptions of the temple found in the Bible, has been discovered at an archaeological site elsewhere in Jerusalem.

Sacrificial altars and incense tables from this period have also been found throughout Israel: at Beersheba, Megiddo, and Arad. These attest to the widespread religious practices of the time. While the temple in Jerusalem was the main place of worship, regional sanctuaries existed throughout the land. Worship here was mixed: some was of the single God of the slowly evolving monotheistic faith, while other was of preexisting pagan gods.

We have far more archaeological evidence for the Second Temple and the time of Jesus. A first-century A.D. ossuary found in Jerusalem bearing the inscription "Simon, Builder of the Temple," gives strong evidence for the rebuilding of the Second Temple by King Herod. Another important archaeological find is a stone inscription in Greek that warns: "No foreigner shall enter . . . ," believed to have been placed by the entrance of the temple. Temple vessels from the first century A.D.—including a menorah, an incense altar, and a showbread table—have also been found and are currently displayed at the Israel Museum.

Many decorated ossuaries from the time of Jesus have been discovered in Jerusalem and the surrounding area. At the end of the Second Temple era, an unusual practice became prevalent here, eventually spreading throughout the region. When a person died, the body was buried in a pit inside the family tomb. After a year, the flesh having all decayed, the bones were collected and placed in the ossuary. Some experts have suggested that this custom reflects the belief in the resurrection of the dead at the End of Days. Ossuaries were made of soft limestone and carved with decorative motifs. One of the most stunning ossuaries ever found, now displayed in the Israel Museum in Jerusalem, bears an inscription in Aramaic, "Jesus son of Joseph," though its authenticity has been disputed. Did it once contain the bones of Jesus of Nazareth?

In addition, an ossuary bearing the inscription "Joseph son of Caiaphas" has been discovered. This ossuary is believed to have held the bones of Caiaphas, the high priest at the temple during the time of Jesus. Also, an inscription in Latin bearing the name of Pontius Pilate was discovered at the ancient site of Caesarea, where the Roman governor resided. It was uncovered near the site of the Roman amphitheater of this ancient coastal city by Italian archaeologist Antonio Frova in 1961; news of the find caused a stir, as it is the only authenticated piece of archaeological evidence for the life of Pilate. Taken together, these archaeological finds support the historical accuracy of some of the general events described in the New Testament.

Evidence of at least one crucifixion has been discovered. This last find was an anklebone with a large iron nail driven through it. It was found in an ossuary discovered in the north of Jerusalem, inscribed "Yehohanan son of Hagkol/Hezkil." The Roman practice of crucifixion was followed and it is believed that many thousands found their end in this terrible way for crimes ranging from stealing to murder.

While the archaeological record of the events of the Bible is not perfect, as time goes on, more is being found that is consistent with the biblical narrative. What has been described in this chapter is only a small sample. Archaeological museums hold troves of finds that have been unearthed over the years.

For proof of the existence of the Hebrew dynasty one need not even visit the Middle East. Titus's Arch, in Rome's Forum, bears a perfectly preserved relief of Jewish slaves carrying the menorah (votive candelabra) from the Temple of Yahweh in Jerusalem (the Second Temple) when it was destroyed by Titus, son of the emperor Vespatian, in A.D. 70. A visit to Jerusalem would offer further proof: the Burnt House, near the temple ruins, has been shown to have been the home of one of the prominent priests, and it is well excavated and preserved.

The Dead Sea Scrolls, the first of which were discovered by an Arab shepherd who stumbled into a cave at Qumran, in the Judean Desert, in 1947, and others later excavated throughout the 1950s, are among the most important archaeological finds in history. These scrolls, most of them made of parchment but some of papyrus, provide evidence for some of the narrative of early Judaism and the time of the emergence of Christianity. The scrolls include most of the books of the Hebrew Bible with very small differences between their text and that of later versions.

The Dead Sea Scrolls date from the third century B.C. to the first century A.D. The finds also include books now considered within the Apocrypha, thus outside the canon, as well as interpretive texts. These scrolls bear evidence for some of the stories of the Bible, dating to twenty-one hundred years ago.

The first seven scrolls were discovered by Bedouin of the Ta'amra tribe in late 1946 and early 1947 in a deep cave by the northwestern shore of the Dead Sea. A short while later, these scrolls were sold by the Bedouin to two antiquities dealers in the city of Bethlehem. The dealers went on to sell the scrolls to others and eventually four of them ended up in the possession of Athanasius Samuel of the Syrian Orthodox Church in Jerusalem. The three other scrolls were bought by Professor Eliezer Lipa Sukenik on behalf of the Hebrew University of Jerusalem.

Because of the unstable political situation in the Middle East at the time, with war looming on the horizon, Athanasius Samuel smuggled his four scrolls into the United States in 1948 and exhibited them over the next three years at various libraries, universities, and art dealerships, hoping to attract buyers. He could find none, so in 1954 he advertised the scrolls in the *Wall Street Journal*. Finally, he got a response: a certain Yigael Yadin contacted him and immediately paid the high price Samuel had asked for. Yadin was secretly acting on behalf of the state of Israel. In fact, Yadin, who would become Israel's premier archaeologist, was the son of Professor Sukenik of the Hebrew University. The four scrolls were thus reunited with the remaining three in the possession of the Hebrew University. They are now displayed in the Shrine of the Book at the Israel Museum in Jerusalem.

In the years 1949 to 1956, further excavations and searches

were conducted of the area of Qumran, and these led to the discovery of a total of 930 new scrolls. All were found within a range of less than two miles of the original site of discovery at Qumran. Most of the scrolls are written in Hebrew, but some are in Aramaic and others in Greek. Out of these scrolls, only twelve are in a state of good preservation; the rest are fragmentary, though scholars have been using a variety of tools over the years to try to piece together the incomplete writings.

The only two books of the Old Testament that are absent from the finds of Qumran are Nehemiah and Esther. The fact that all other biblical books are found in these scrolls lends strong support for the belief that the biblical texts were studied during the time of Jesus and earlier. The Qumran sect, whose members wrote, studied, and preserved these unique documents, seems to have thrived for almost four centuries. When Jerusalem was destroyed by the Romans in A.D. 70, following the Jewish Revolt that had begun in A.D. 66, the Romans proceeded to the south and within three years caught or killed all members of the sect who had not managed to escape. Before the arrival of the Romans, the sect members hid their texts in deep caves in the area to preserve them for posterity. And it was thus that they were discovered in the middle of the twentieth century and have brought us, in addition to texts of the Old Testament, documents with a wealth of information on the life of a mysterious ancient group that lived in biblical times.

We have seen in this chapter that historical and archaeological evidence exists for parts of the general biblical narrative. There is, of course, no archaeological proof for miracles or supernatural acts; but the general historical events recounted in the Bible are in agreement with the growing archaeological record of the Holy Land.

# 3

## The Revolt of Science

O ver the centuries, religion assumed a moral and spiritual
role in human life and also began to rely more and more
on the supernatural. This hampered advances in natural science for many centuries. The Greek civilization had advanced ideas about the natural world—including Democritus's
perceptive notion of atoms and the heliocentric view of the
solar system advanced by the fourth-century b.c. philosopher
Philolaus. When this great culture declined and the Western
world sank into the Dark Ages, Scripture assumed the role
of the explanation of truth, and freethinking was shunned.
This mode of thought continued through the late Middle
Ages, when except for the development of crude notions about
medicine (most of them wrong, such as prescribing bleeding
as treatment for various ailments) there were few attempts to

pursue science. Deviations from established belief were not tolerated in a culture dominated by the church and Catholic monarchs. Simply put, the "order of things" was not up for debate.

This logic extended to geocentrism. The belief that the Earth was the center of the universe was a key, entrenched principle. This theory had found its "scientific" explanation for the movements of all visible bodies in our solar system not in the Bible, but through the work of Claudius Ptolemy (c. A.D. 90–c. A.D. 168), an astronomer and mathematician who lived in Alexandria during the long twilight of classical Greek civilization. In his treatise the *Almagest* ("The Great Work"), Ptolemy outlined a model of the solar system that maintained Earth as the center of creation and yet explained mathematically all the motions of the planets and the moon, as well as the (apparent) motion of the sun around the Earth. By the Middle Ages, the "Ptolemaic worldview" had become entrenched, as much a part of dogma as the Gospels.

How did Ptolemy's system work? Because planets exhibit retrograde motion in the sky—meaning that at times a planet will seem to move backward, due to our planet passing it in our orbit around the Sun—Ptolemy posited *epicycles* (cycles within cycles). Thus the sun rotated directly around the Earth, but while the planets also orbited the Earth, they simultaneously rotated in another, smaller circle whose axis was on the main orbital path. In this way Ptolemy accounted for the retrograde motion of planets as they appear in the night sky.

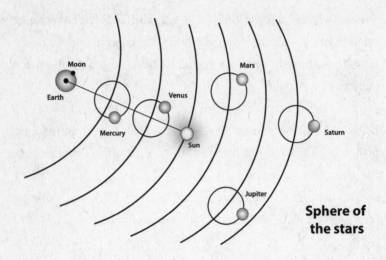

*Ptolemy's erroneous model of the solar system, requiring the use of
complicated epicycles to explain the motions of the planets.*

What is so interesting here is that this model works! It
"explains" all the motions of the heavenly bodies around us.
Unfortunately, it is completely wrong. The Ptolemaic system
demonstrates that we can concoct very complicated systems
that seem to explain a phenomenon. The true model in the case
of our solar system is, of course, the one devised in the mid-
sixteenth century by the Polish astronomer, mathematician, and
jurist Nicolaus Copernicus (1473–1543).

What are the differences between these two mathematical
explanations of the solar system? Besides the fact that direct

*The Copernican model of our solar system, much simpler than Ptolemy's and correctly describing planetary motions.*

observation confirms the sun is at the center of the solar system and everything else moves around it, the Copernican model is *simpler* and more elegant—it requires fewer assumptions. According to the principle of Occam's razor, the simplest explanation of a phenomenon is probably the correct one. A good model explains much about the world without employing too many special assumptions—such as the epicycles—and without requiring too many complex notions and tools. Einstein would make an art of creating this kind of model; he once said, "A scientific theory should be as simple as possible, but not simpler." Everything essential should be in the model, but nothing too elaborate or overly complicated unless it serves a clear purpose. Einstein's theories of special and general relativity are models of genius. They are not simple, but every element in them is absolutely essential. They are parsimonious: every symbol, every equation, every element is exactly what it needs to be in order to explain the world.

Copernicus's theory spelled the end of the Ptolemaic model of the solar system, which dictated how people thought about the cosmos for a millennium and a half. His model was far simpler and explained the motions of the planets and the moon, and the apparent movement of the sun through the sky, far better than Ptolemy had done.

**IN 1543, THE** year he died, Copernicus published his landmark book, *De Revolutionibus Orbium Coelestium* ("On the Revolutions of the Celestial Spheres"). Copernicus had studied at Kraków and later at the University of Bologna—the oldest university in the world, founded in 1088—as well as in Padua. In Padua he conducted astronomical observations of stars and planets and studied the works of Greek mathematicians and philosophers, paying special attention to the theory of Ptolemy. He became progressively more certain that the Ptolemaic model of the solar system could not possibly be right. A much simpler model that reproduced all the astronomical evidence in a more concise and natural way was one that placed the sun at the center of the system.

Copernicus wrote a forty-page commentary explaining his heliocentric model. Then, at the urging of his pupil the Austrian mathematician Georg Joachim Rheticus, he agreed to publish his theory in a book. Thus *De Revolutionibus* was born. Unfortunately, Copernicus died just as the book saw publication—he is reputed to have held the first copy in his hands as he died peacefully in his bed. He likely had no inkling that his book would mark the beginning of modern science.

Copernicus did not intend to antagonize religious people. He simply understood that the sun was the center of the solar system, and he wanted everyone to know it. He had close relations, through family members, to Catholic officials, and the book's publication was helped by a Lutheran cleric, Andreas Osiander.

But *De Revolutionibus* was mostly ignored when it appeared. Far from igniting an immediate conflict between religion and science, it slipped under the radar; and in any case, its author had died. Copernicus's revolutionary idea couldn't remain hidden, however, once another genius got hold of it.

**PERHAPS MORE THAN** any other scientist in history, Galileo Galilei (1564–1642), the Renaissance genius who was responsible for so many great discoveries about the physical universe, epitomizes the sharp conflict that erupted between religion and science. Galileo, too, was far from antireligious. In fact, one of his daughters was a nun, and he was never far from religious circles through acquaintanceships and friendships—even the pope, knowledgeable about mathematics, was sympathetic to him.

Galileo possessed one of the most penetrating, sharp, and curious minds in history, and his lifelong quest was to decipher nature's laws, regardless of what people, books, the Bible, or the church might tell him to believe about nature. Galileo famously experimented with falling and rolling objects, perhaps even dropping them from the top of the Leaning Tower of his native Pisa.

Galileo's early experiments easily debunked the centuries-old belief, originating with Aristotle, that objects fall at speeds proportional to their weights. In fact, in the absence of air resistance, the proverbial feather would fall to the ground at the same time as a weight of lead (air resistance will slow down a feather considerably, which is how birds can fly in the first place).

In 1609, soon after the telescope was invented in Holland, Galileo obtained one, then fashioned some of his own. At first, he aimed to sell these new devices to the fiercely independent Venetian republic, to help its naval leaders defend against possible attacks from the sea by using a telescope to scan the horizon from atop the campanile in St. Mark's Square. But the invention didn't sell well, and Galileo began turning his telescope to the night sky. Soon he made one of the most important discoveries in history, earning him recognition as the father of astronomy. (He is also known as the father of modern science.)

On January 7, 1610, Galileo made observations of Jupiter and discovered four of the giant planet's satellites, now named after him as the Galilean moons (Ganymede, Callisto, Europa, and Io). Here were heavenly bodies that clearly orbited an entity other than the Earth, as Galileo concluded over several nights of observation, seeing the moons change sides as they traveled around the planet. This discovery clearly contradicted the Catholic Church's Ptolemaic belief that every celestial body orbits the Earth.

But the coup de grâce came within eight months, in September of 1610, when Galileo observed the planet Venus and noted that it went through complete phases, like the moon. According to the Ptolemaic model, built on epicycles, we should only be able to see some of Venus's phases: either thin crescents (if it were on the inside of the orbit of the sun around the earth) or only gibbous and full phases (if it were on the outside of the sun's orbit)—but not both. The fact that all phases of Venus are visible, just as we see with the moon, meant to Galileo that the Ptolemaic model couldn't possibly be right. Rather, Copernicus's model of the solar system was the one that reflected reality.

Galileo began publishing his findings in books that irritated the Church. He kept progressively running afoul of the Roman Inquisition, especially when his book *Dialogue Concerning the Two Chief World Systems* appeared in 1632, in which he mocked the church and its advocacy of the geocentric model by expressing these opinions through the mouth of Simplicius, the simpleton in the dialogue. Galileo was offered refuge from the church in independent Venice but chose to remain in his native Tuscany, whose dukes were more closely allied with and influenced by Rome.

Eventually, the Tuscan rulers could not protect Galileo from an order of extradition to face the dreaded Inquisition in Rome—which had already put to death many independent thinkers for ideas contrary to the official Catholic understanding of the universe. Although Pope Urban VIII was sympathetic to him, even the pope could not deter the Inquisition in its persecution of Galileo. The infamous trial took place in February 1633. Under threat of torture, Galileo publicly recanted his heliocentric heresy, perhaps muttering sotto voce *"Eppur si muove"* ("Still, it moves!") and was confined for the rest of his days to house arrest in his villa at Arcetri, outside Florence. Galileo's trial, more than any other event in history, has come to symbolize the sharp split between science and faith, a conflict that continues to rage (to a degree, and in very different form) in our own time.

Galileo achieved another landmark: he forged the immensely important alliance between mathematics and science. Galileo's famous statement "The book of nature is written in the language of mathematics" would chart a common course for physics and mathematics that would continue and intensify into our own time and well into the foreseeable future.

But Galileo also made a key discovery in pure mathematics. Confined to his Arcetri estate late in his life, Galileo thought about infinity and recognized that infinite quantities possess very strange qualities. Galileo considered the infinite set of all positive integers (1, 2, 3, 4, 5 . . .) and the infinite set of all *squared* integers (1, 4, 9, 16, 25 . . .). Both sets are infinite, but one could draw a one-to-one correspondence between each member of the first set with a unique member of the other set. Thus, Galileo matched 1 with 1, 2 with 4, 3 with 9, 4 with 16, 5 with 25, and so on. So while there are infinitely many numbers in each set, and while the set of squared integers is clearly a proper subset of all the integers (since the integers that are squares are also members of the set of integers), the one-to-one matching of every integer with every squared integer shows that there is "the same number" of integers as of squared integers. We say that the two infinite sets have the same *cardinality,* the same "size." We will come back to this discovery later in the book.

Following Galileo's immensely important astronomical discoveries, it became impossible for serious astronomers to maintain a geocentric view, and so a middle ground was reached in which, in order not to antagonize the church, the planets were seen to go around the sun, but the sun and its planets rotated around the Earth. This hybrid model was the one maintained by the Danish astronomer Tycho Brahe, who made a large number of stellar and planetary observations, first at Hven, an island on which the king of Denmark gave him land for his observatory, and later in Prague, where he moved to continue his work.

The massive set of data on the positions of stars and planets collected by Brahe was used by his assistant, the brilliant German mathematician Johannes Kepler, to derive his laws of planetary motion—in a fully heliocentric model. Kepler's laws are so accurate that they are still used today, four centuries later, in astronomical work, including determining the orbits of newly discovered extrasolar planets as well as directing spacecraft to approach, orbit, and land on planets in our solar system.

Just as Copernicus and Galileo did not intentionally seek to break science away from religion and remained close to people of faith throughout their lives, neither did Kepler pursue his scientific work with an intention to challenge faith. Rather, it was the Catholic Church, with its strictly literalist approach to Scripture and its centuries-long adherence to Aristotelian philosophy, that was the belligerent side in this deepening dispute.

Kepler's work presaged the invention of the calculus by Newton and Leibniz in the following century. In science, he was both an astronomer and an astrologer—in the sixteenth and seventeenth centuries, science and the occult remained in a somewhat nebulous common milieu. Similar mixing of science, spirituality, and the occult is found in the work of René Descartes, who contributed much to mathematics, science, and philosophy in the seventeenth century.

**DESCARTES WAS BORN** on March 31, 1596, to a wealthy family belonging to the French aristocracy. Though he was born in the town of La Haye (now renamed Descartes in his honor), in the region of Touraine, in western France, the family lived in the neighboring region of Poitou. He spent his childhood moving between both regions.

Touraine and Poitou were very different. While Poitou was mostly Protestant, Touraine was predominantly Catholic, as is most of France even today. This exposure to both sides of a religious conflict affected Descartes's feelings about religion and society throughout his life. One key example of this was his excessive fear of the Inquisition and of what it might do to him if he published scientific writings contrary to church doctrine, complemented by an almost naive lack of concern about igniting Protestant ire.

Descartes studied at the Jesuit College of La Flèche in western-central France. He was far from antireligious. His closest friend throughout his life was the Minim monk Marin Mersenne, a cleric equally interested in mathematics and science whom he met while both were students at La Flèche. According to his biographer Stephen Gaukroger, Descartes remained a believing Catholic until the end of his life.

Descartes traveled widely, often joining military campaigns as a volunteer soldier. He participated in the siege of Prague and other campaigns—employed by either Protestants fighting Catholics or vice versa during the infamous Thirty Years' War. During one of his travels in southern Germany, he met the mathematician and mystic Johann Faulhaber. Faulhaber showed Descartes some of his work on equations, and as a consequence Descartes later adopted Faulhaber's mystical notation.

For example, he used the sign of Jupiter ♃ in his algebraic manipulations.

The philosopher and mathematician Blaise Pascal, who along with the mathematician Pierre de Fermat formulated the theory of probability, was a friend of Descartes. Pascal contributed greatly to physics and mathematics, and was a deeply religious man. He is now best known for conceiving Pascal's wager, the famous rationale for believing in God on the (somewhat cynical) grounds that to choose otherwise is far too risky: if you believe, and God doesn't exist, according to Pascal's logic you pay no price (or at least no great price); but if you do not believe in God and he does exist, then the cost to you is eternal damnation.

Both Fermat and Descartes read the works of Euclid, translated into Latin, and learned from them about the wisdom of the ancient Greeks. Equally, they assiduously studied the works of the great Galileo. The two extended the physical-mathematical learning of the ancient Greeks, as well as of their own near-contemporaries Copernicus, Kepler, and Galileo, to the new setting of the seventeenth century in which physical science, hand in hand with mathematics, would forge new paths.

As he studied the physical world, Descartes became obsessed with the Inquisition. He was aware of the travails of Galileo and was determined not to encounter a similar fate. In letter after letter to his friend Marin Mersenne, the monk, he expressed his fear that if he published a work proving that the Earth rotated around the sun, rather than the opposite view supported by church doctrine, he would be hunted down by the Inquisition. He had written a book called *Le Monde* ("The World") but refused to publish it since it espoused such views. To further protect himself, during his travels Descartes kept in touch only with Mersenne, and would usually post his letters from nearby places, not the actual locations where he was at any given time.

Eventually, in 1628, fear of the Inquisition made Descartes decide to move to Holland, even though the king of France had given him special privileges as a prominent philosopher and scientist. Becoming even more obsessed with the fate of Galileo after receiving news of his 1633 trial by the Inquisition, Descartes kept changing addresses, living in various little towns in Holland.

In 1637, Descartes finally published his masterpiece, *The Discourse on the Method*, which contained the essence of his philosophy. This classic book had an appendix called *La Géométrie*, which detailed Descartes's breakthrough work to wed geometry with algebra and introduced the essential idea of the Cartesian coordinate system.

His ideas made Descartes immensely famous all over Europe—but they also attracted new enemies. In 1647, Descartes was accused by Protestant theologians of being an atheist, which he certainly was not. When Descartes defended his views in print, he was subsequently accused of libel. A court decided against Descartes, and he was made to write a humiliating letter of apology to the theologian who had attacked him.

Deeply hurt, he decided to move to Sweden, where in 1650 he died from the flu—or perhaps was poisoned by religious opponents; the cause of his death is still a mystery. After his death, Queen Christina of Sweden, citing Descartes's influence, converted to Catholicism.

Descartes was a brilliant mathematician and physical scientist who maintained his faith in God and saw no internal conflict between science and religion, despite his lifelong difficulties. Descartes took the physical investigations of Galileo to a whole new level. He understood that the Earth rotated, and he knew that we are not the center of the universe. He thus questioned the literalism of the church and understood that reality did not agree with church beliefs. Descartes's groundbreaking scientific ideas were pursued by later scholars.

BY AGE TWELVE, the German philosopher, statesman, and mathematician Gottfried Wilhelm Leibniz (1646–1716) was fluent in ancient Greek, having deciphered the language on his own by reading Plato and Aristotle, which introduced him to logic and inspired a passion for the foundations of pure reason. At the same time, he also developed an interest in theology.

In 1661, Leibniz enrolled at the University of Leipzig, where he studied the works of his contemporaries Hobbes, Bacon, and Galileo. In 1663, he presented his thesis, "De Principio Individui," which dealt with the ideas of individuals and totality. This led to his concept of a *monad*. Leibniz's quest was to answer a major question that goes back to the ancient Greeks:

What is space? This leads to other questions that relate to it: What are points? What are lines? What are planes? What are three-dimensional objects? The monads are an attempt to answer these questions by defining in an abstract way the basic Greek element of space. A monad, like a point in Greek geometry, is something that has no interior: no length, no breadth, no height, nothing inside it. But it is more abstract and also extends into metaphysics by also defining *ideas* in a very general way. The monad is the most abstract basic element of everything in *both* the physical and the spiritual world.

A religious Protestant who was influenced by his association with Catholic princes, Leibniz was attracted to the idea of reconciling the religions of Europe as a way of unifying all the people of the continent. In 1668, Leibniz wrote a treatise arguing for the existence of God and the immortality of the soul. Titled "Nature's Testimony Against the Atheists," it was grounded in his grand scheme of uniting Europe's warring religions.

Leibniz invented the calculus, independently of Newton. But he viewed all his work within the context of a whole. He strove to apply his new mathematics to theology. The infinitesimals he invented for his work on the calculus—or rather adapted from the works of the ancient Greeks—held mystical powers in his eyes, and he hoped to be able to use them also in metaphysical investigations. While Newton, too, was a religious man, his calculus was steeped in the needs of physics rather than metaphys-

mately, the theory of calculus independently developed by Leibniz and Newton would constitute one of the main tools physicists would use in their investigations of nature and their attempts to uncover its ultimate laws.

**ISAAC NEWTON (1642–1727)** was born on Christmas Day of the year Galileo died. He came from a farming family in Woolsthorpe, in the county of Lincoln, England. Newton was a premature baby, and his mother once described him as having been so small at birth that he could fit inside a quart mug.

To describe his greatest scientific achievements, Newton famously said, much later, "If I have seen a little farther than others, it is because I have stood on the shoulders of giants." Presumably, the giants on whose work the modest Newton had relied included Descartes, Kepler, and Galileo. Cartesian logic led him forward, and Descartes's ideas had verged on the full calculus. Galileo's investigations of falling bodies and other physical phenomena inspired his work in physics. And Kepler's laws of planetary motion were logical corollaries of Newton's law of universal gravitation. Newton was not a man of interests as wide-ranging as those of his contemporary Leibniz, but in the realms of physics and mathematics, Newton's intellect was supreme.

In 1664, Britain suffered an attack of bubonic plague, and Cambridge University, where he was a student at that time, closed down. Newton left for Woolsthorpe, where he spent two years alone, meditating on the universe and its laws. It was here that Newton invented the calculus. He called it the method of fluxions (the word comes from the idea of a flow). Newton viewed variables as flowing, and to describe this movement—the rate of change of a quantity with time—he invented the differential calculus.

Newton's law of universal gravitation states that any two particles of matter attract each other gravitationally with a force that is proportional to the product of their two masses and inversely proportional to the square of the distance between them. The constant of proportionality in the equation is known as Newton's constant, $G$. Newton also developed laws of motion, which state that every body will continue in its state of rest or inertial motion unless it is acted on by a force; that the rate of change of momentum (which is mass times velocity, in Newtonian physics) is proportional to the force acting on a body; and that action and reaction are opposite and equal to one another: for every action there is an equal and opposite reaction.

Newton was a religious Unitarian, and he also spent much time during his two years in Woolsthorpe trying to make sense of the predictions of the biblical prophet Daniel as well as understanding the Apocalypse. Attracted to big ideas, he saw no conflict between pursuing groundbreaking science and at the same time contemplating the most difficult religious ideas.

While by Newton's time a serious rift was evident between the teachings of the Catholic Church and the pursuit of science, many of the major scientists of the day remained devoutly religious people. This may seem paradoxical, but the truth is that the fissure that had materialized between science and faith was not created by the scientists, but rather by the church. Its most dramatic manifestation was the trial of Galileo, though that was far from the only act of persecution of scientists and thinkers by the Inquisition. But in the nineteenth century, science would succeed in proving many assertions that the church viewed as contrary to the Bible.

With Newton's important achievements—building, of course, on the work of the great thinkers who preceded him—civilization reached an extremely high level of knowledge about the universe. Newtonian mechanics and also the optics, astronomy, and mathematics to which Newton contributed were immensely valuable in helping us gain an understanding of the vast and complicated physical world around us. The progress made by Newton and others allowed us to see clearly how planets move and how the force of gravity operates in nature. Newton's theory of gravity is so profound and so all-encompassing that its laws govern everything from the fall of apples to the ground to the orbiting of the Earth by the moon; from the revolutions of the planets around the sun to the actions of springs and the trajectories of cannonballs; from the behavior of billiard balls to the energy of an accelerating car. Newtonian mechanics explains the world to a stunningly accurate degree. In the twentieth century, Einstein would refine Newton's theories to account for instances where the speed of light is approached or very great mass is involved. But for the time, Newton's achievements truly opened a new world for physical science. In the following two centuries, science would consolidate its gains, and the church would have to retreat from its position as the source of truth about how the world works.

# 4

## The Triumphs of Science in the Nineteenth Century

Science and religion gradually went their separate ways over time, and in the nineteenth century, science won a number of key victories against religion—at least against some of the more unjustified beliefs about the universe associated with the Judeo-Christian tradition. This was, perhaps, the last time in history that science could easily dispel views of the world encouraged by organized religion.

**BOTH RICHARD DAWKINS** and Christopher Hitchens quote an exchange between the emperor Napoleon and the brilliant French mathematician Pierre Simon de Laplace. Laplace had taken Newton's work and extended it in a substantial way to the entire solar system, presenting his results in *Celestial Mechanics,* first published in 1799. Newtonian physics was thus shown to govern the complicated interactions among the planets. Some time later, Laplace made a present of his book to Napoleon. The emperor read it and then said to Laplace: "You have written an entire book about the world without mentioning its maker." Laplace's response was: "Sire, I had no need of that hypothesis."

Unfortunately, both Dawkins and Hitchens leave out the final punch line of this story: Napoleon later repeated his exchange with Laplace to another great French mathematician, Joseph-Louis Lagrange, who had also done important work in astronomy and mathematics. Lagrange's answer was *"Ah! mais c'est une belle hypothèse; ça explique beaucoup de choses"* ("Ah, but it is such a beautiful hypothesis; it explains many things"). For our purposes, the point here is that there were other views even among the greatest mathematicians and astronomers at that time.

Science in general was making great strides at the time of Laplace and Lagrange, as the eighteenth century gave way to the nineteenth. Still in the eighteenth century, the Scottish geologist, chemist, medical doctor, and naturalist James Hutton explained geological processes taking place in the Earth; his work indicated that there were forces at work that operate in *geological time,* a far longer range than the few thousand years commonly speculated as the age of the Earth based on biblical chronology. Rock formations and erosion processes, such as those evident in the terrain of the Scottish Highlands, had directed Hutton to a theory of geology that would lead the way to our modern understanding of how Earth processes work.

Almost two centuries earlier, in contrast, the Irish clergyman James Ussher had used biblical chronology and genealogy—by calculating the estimated dates of the lifetimes of biblical figures going all the way back to Adam—to conclude that the Earth was created in 4004 B.C. Ussher was one of many biblical interpreters who used Scripture to try to guess the age of the Earth by various means. The ages others arrived at were all around five thousand to ten thousand years. Today we call the belief that our planet came into being a few thousand years ago the young Earth hypothesis.

In 1799, the English surveyor and geologist William Smith drew the first geological map of Britain, which would eventually also help establish that the Earth was far older than interpreters of Scripture had guessed. Smith's analysis of geological strata was used by his nephew John Phillips in the mid-1800s, along with a study of ancient fossils, to estimate that the Earth was about 96 million years old.

Some thirty years later, William Thompson, Lord Kelvin, concluded that the Earth must be hundreds of millions of years old, based on the physical properties of rocks and geological layers and an estimate of how many years a body the size of the Earth would need to cool down from a molten rock phase to the solid body of matter we see today. The oldest rocks on Earth have now been dated, using uranium-isotope dating, to 4.5 billion years ago, which is now the scientifically accepted age not only of the Earth but of our entire solar system.

Work in geology would inspire another revolution in the nineteenth century: Darwinian evolution. In the meantime, the young Earth hypothesis died under the weight of geological evidence and reasoning (though some literalist Christians continue to believe in it).

Simply walking the Earth, one can see that our planet could not have formed five thousand, ten thousand, or even a million years ago. Digging around on inland hills, one can find shells and other remnants of ancient sea organisms on very high ground. And for a sea to recede so far, or for the earth's elevation to rise so that the former sea floor is in the mountains, takes millions of years.

Similarly, older mountains such as the Appalachians show the effects of erosion through wind, rain, and snow that requires millions of years, as compared with younger mountains such as the Alps, with their jagged edges, cliffs, and bluffs, which show far less extensive damage through erosion processes. Mountain chains rise, earthquakes take place along fault lines, plates move under one another forcing mountains to move upward and volcanoes to erupt—all over millions of years. Walking in the Alps, Alaska, or Ecuador's Cordillera Blanca, one can clearly see the moraines that were carved by glaciers and now appear as U-shaped valleys, formed when the glaciers receded over very long time periods. Fossils of long-gone species that once roamed the Earth, including the dinosaurs, help seal the fate of the young Earth hypothesis.

**THE BIBLICAL STORY** of creation in six days thus is a literary device. Needless to say, counting biblical lifetimes back to Adam is not a good way to estimate our planet's age. Scripture is written allegorically, not literally, and it does not agree with what scientific observation clearly tells us. The same is true of the literal idea that the sun rises and sets around a static Earth.

If you stay up and look at the night sky for a long time, it will become clear that the Earth rotates and not the sky. Why? Because the stars and constellations move from east to west at a *uniform rate*. If you were in a moving car but didn't know it, and through the window you saw that all the trees by the side of the road were receding from you at the same speed, you would quickly know that it was you who was doing the moving. Clearly it is not the case that all the trees had suddenly decided to run away from you at the same speed. The same is true of the stars you see on a single night of observation. Planets sometimes transverse the sky "backwards"—the retrograde motion—but this is not evident during a single night. So what you watch in a night—and if you leave a camera on for even a few minutes, the streaks of star motions will verify it—must mean that we are moving, not the entire sky transporting itself around us.

In 1851, an untrained, self-educated French physicist named Léon Foucault dealt the final blow to the belief in a stationary Earth. Astronomers and many educated people had understood well by then that it was the Earth that was doing the rotating. But this notion was not strong enough to provide proof that would satisfy *everyone*. People were awaiting a *terrestrial proof* that indeed it is our planet that is rotating, as if saying: "Show us that we move!"

Foucault was a keen observer and experimentalist who had made many inventions and even measured the speed of light to some accuracy. Thinking about the rotation of the Earth convinced him that he might be able to provide the desired proof if he could only hang a pendulum "above the Earth"— meaning attached to the ceiling of a room in a way that its plane of swing would remain free to move and rotate around. He knew that if he could show that a pendulum swung in its own plane, independently of the Earth, this plane of swing would seem to us on the ground to rotate in a circle. With such a demonstration he would be able to prove that the Earth moved.

Foucault never married and lived with his mother. She was a wealthy woman and owned a house at the intersection of rue d'Assas and rue de Vaugirard in the fashionable Sixth Arrondissement on the Left Bank in Paris (an elegant plaque at this intersection marks the location of Foucault's house). Foucault used the basement of his mother's house for his laboratory. He designed and hung an apparatus on the ceiling of the basement that would allow for free, frictionless rotation of a pendulum that was hung from it. Foucault was not only a great scientist and engineer, but also had artistic skills. He made an attractive copper bob for his pendulum, which can be admired today at the Musée des Arts et Métiers in Paris.

On January 6, 1851, at exactly 2 A.M., after experimenting with his pendulum for hours, Foucault finally saw what he was looking for. He suddenly noticed that the Earth was slowly rotating right under his pendulum. There followed later experiments, conducted in public: first at the Observatory of Paris, to which scientists and savants were invited; and later in the greatest public building in Paris: the Pantheon, to which the French aristocracy, including Napoleon III, were invited. An acute observer both of the physical universe and of human nature, Foucault described the effects of his pendulum experiment as follows:

> The phenomenon develops calmly, but it is inevitable, unstoppable. One feels, one sees it born and grow steadily; and it is not in one's power to either hasten it or slow it down. Any person, brought into the presence of this fact, stops for a few moments and remains pensive and silent; and then generally leaves, carrying with him forever a sharper, keener sense of our incessant motion through space.

*A modern reconstruction of Foucault's pendulum experiment, which finally proved definitively that the Earth rotates, in Paris in 1851.*

The 1851 demonstration in the Pantheon was decisively convincing, and the experiment has since been successfully repeated at many different locations around the world since. In 1913, the Catholic Church finally accepted Foucault's proof that the Earth, indeed, rotates. But it was not easy, and the fact that it took sixty-two years from the definitive experimental proof for the church to accept that we are living on a moving, rotating planet shows how hard science has had to fight to persevere against scriptural interpretations.

Eight years after Foucault dramatically proved the rotation of the Earth, another pillar of traditional religion was destroyed forever: the literal interpretation of the biblical story of the creation of Adam and Eve. In 1859, Charles Darwin published his book *On the Origin of Species,* which laid out the revolutionary theory called *evolution.* Analyzing fossil records and the results of observations of nature obtained over his five-year voyage aboard the *Beagle* had led Darwin to the key concept of natural selection. (Alfred Russel Wallace had a similar idea some time before Darwin's publication of his theory, and both theories were published around the same time.) Evolution showed the descent of human beings from earlier ancestors and of all species from older and simpler life-forms.

Modern genetic research has elucidated Darwinian evolution and provides many pieces of information in its support. Today evolution is a pillar of modern biological science and explains a host of phenomena. Fossil discoveries in the nineteenth century further made it clear that the Earth was once home to living creatures that are no longer around: mastodon bones uncovered in the Americas and mammoth remains in Europe, for example, made it clear both that the Earth is very old and that species have come and gone over thousands and millions of years. Neanderthal remains found in Europe beginning in the nineteenth century led to the understanding that other human species once inhabited our planet, left their mark—including amazingly efficient stone tools used to kill and butcher animals—and disappeared as our own species swept through Eurasia between forty thousand and thirty thousand years ago.

The discoveries of fossils, evolution, geological time, the rotation of the Earth, and other developments in science showed that Scripture should not be taken literally. It is important to note, however, that when the literal interpretation of the Bible is abandoned, Scripture does not necessarily seem "wrong." For example, the order of appearance of living things on Earth described in Genesis is not in disagreement with evolution: lower forms come first, starting with flora, then lower fauna, and then more advanced animals, leading to humans. The order is surprisingly correct—they are just not "created" from one day to the next.

# 5

# Einstein, God, and the Big Bang

The Bible talks about the creation of the world. Cosmology is the branch of science that addresses the inception of the universe, making use of the theories of relativity and quantum mechanics. In this chapter we look at how science and religion view the birth of the cosmos.

Modern theoretical physics began in 1905, when Albert Einstein first shattered our conceptions of time, distance, and speed in his special theory of relativity. Einstein's theory showed that time is not constant and that both time and space must "bend," contract or expand, to accommodate the universal constant—the speed of light. Nothing goes faster than light; as you travel closer and closer to the speed of light, your mass increases toward infinity and time slows down for you (as measured by a stationary observer).

Einstein also taught us in 1905 that mass is equivalent to

energy, using his famous equation $E = mc^2$. This principle is exploited in the Large Hadron Collider at the international physics lab called CERN (Conseil Européen pour la Recherche Nucléaire, the European Organization for Nuclear Research), near Geneva, Switzerland. This is a giant particle accelerator that operates on the principle that the effective mass of a highly accelerated particle becomes very large, thus generating huge amounts of energy when the particle crashes.

In turn, this energy changes into new particles that had never before been seen, such as the famous "God particle," the Higgs boson, whose discovery has recently been announced. This particle is believed to have been present shortly after the Big Bang and to have given mass to itself and to other massive particles in the universe.

According to our latest cosmological theories, the Big Bang created only sheer energy. Particles with mass, such as electrons and quarks—which make up protons and neutrons and thus the nuclei of all matter—received their mass from the Higgs boson. One particle remained massless: the photon, the ubiquitous particle of light.

It's important to understand that the experiments in the Large Hadron Collider, which mimic what happened in the Big Bang, do not create anything "out of nothing." The energy is given to the protons by accelerating them in a tunnel using strong electric fields, and intense magnetic fields keep the particles going in circles. The entire operation consumes electrical energy equivalent to that used by an entire city.

Existing particles are given kinetic (motion) energy through electric energy turned into electromagnetic fields; when they crash they release exactly the energy they obtained (plus their rest energy, determined by Einstein's formula), and this energy produces new particles. But the entire operation actually demonstrates one of the most important principles in physics, called energy conservation: energy (in the form of mass, radiation, or motion) simply changes form—it can be neither *created* nor *destroyed*. Nothing here comes "out of nothing."

Einstein realized that the great breakthrough he had achieved with his special theory of relativity would affect gravity and how it acts. He knew that the relativity principle should alter Newton's great edifice, the theory of mechanics. When objects move very fast, or when their masses become very large, corrections to Newtonian mechanics would have to be made.

A general theory of gravity that would incorporate Newton's work with his own required Einstein to immerse himself in high-level mathematics for several years. Finally, in late 1915, ten years after he presented his special theory of relativity, Einstein had a complete relativistic theory of gravity: the general theory of relativity. He published it in early 1916. Einstein's equations of general relativity enjoy mathematically pleasing properties of symmetry, structure, and what both mathematicians and theoretical physicists like to call "elegance." The equations are concise and precise: they contain everything that is necessary for a model of a complicated physical system and nothing superfluous—in Einstein's words, "as simple as possible, but not simpler"—and they achieved his goal of explaining gravitation in a relativistic way.

Einstein's general theory indicates that space curves around massive objects: it is in this sense a geometric theory, since it shows that the geometry of space-time changes with the effects of gravity. Massive objects curve space around them. Space and time themselves are interlinked in what we call space-time.

Einstein needed physical proof of his theory, which was eventually provided by the English astronomer and physicist Arthur Eddington, the secretary of the Royal Astronomical Society. During World War I, papers from the enemy were not allowed into Britain, so Einstein arranged to have a friend in Holland, the physicist Willem de Sitter, send a series of Einstein's papers to Eddington. Eddington thus became aware of general relativity long before anyone else outside Germany.

Eddington was a conscientious objector who refused to fight in World War I. Because he was already a renowned scientist, who had developed important theories on processes that take place inside stars, Britain allowed him to perform alternate, nonmilitary service to the nation in wartime. His service would be to science: he organized an expedition to the island of Príncipe in the Atlantic, where a total solar eclipse was to take place on May 29, 1919. Another group, also under his overall command, would travel to Brazil to view the same eclipse. Both teams would make observations of starlight just grazing the sun, which was hidden behind the moon during the total eclipse, and look for deviations of the starlight that would confirm the bending of space predicted by Einstein's new theory.

Despite the risks of malaria, snakes, and inclement weather on the island, the mission was ultimately successful: Eddington's team at Príncipe and the one at Sobral, Brazil, provided photographs that showed the bending of starlight around the sun exactly in the amount (within expected statistical accuracy) Einstein's theory had predicted. Once the expedition returned to Britain, Eddington's presentation of his results, which received worldwide attention, resulted in Einstein becoming an instant celebrity.

Since then, general relativity has been confirmed in many settings and its other predictions have also been verified. General relativity solved a mystery about why the perihelion of Mercury (the closest point to the sun in the planet's orbit) shifts—a phenomenon that had remained unaccounted for by Newtonian mechanics. Black holes, which were predicted by general relativity, have been confirmed in space by observing matter that falls into them, releasing X-rays. And a host of other phenomena have also been observed and explained using the new theory, including the gravitational redshift: the fact that the wavelength of light is stretched through the effect of gravity.

Einstein's general theory of relativity changed our understanding of nature. Even before the confirmation of his theory by Eddington, Einstein went about trying to apply what he had just theorized to the universe as a whole: he wanted to construct a general-relativistic model to explain the *entire universe* we see around us—something that to almost anyone else would have seemed an outrageously ambitious undertaking, but not Einstein.

By 1917, Einstein had devised a *cosmological model* of the entire universe. Based on the state of astronomical knowledge at the time, Einstein assumed that the "universe" was simply our Milky Way galaxy (Andromeda, the nearest galaxy to our own, visible even to the naked eye from dark locales, was at that time thought to be just a nebula within our own Milky Way home). But Einstein's equation predicted that the universe could not be static. Since he believed that our galaxy was neither expanding nor contracting, Einstein had to "stop" his theoretical universe from changing through time, and so he added a term to his equation, called "the cosmological constant." He thus had a formula for a universe that was constant and static. It therefore had neither a beginning nor an eventual end.

Einstein maintained the cosmological constant in his equation until the early 1930s, when, on a visit to California, he met Edwin Hubble. The latter told him about his 1929 discovery of the expansion of the universe, based on the movements of faraway galaxies that he, along with Vesto Slipher and Milton Humason, had observed with the one-hundred-inch Mount Wilson telescope. While Hubble may not have understood this at the time, an expanding universe had to have once been very tiny—and hence had to have had a beginning. This beginning is now referred to as the Big Bang.

**ANOTHER STORY OF** the beginning of the universe is, of course, told in the Bible's book of Genesis, written by nonscientists perhaps three thousand years ago: First there was nothing at all, and then God created the universe. The writers of Genesis understood that the cosmos had to have had a *beginning*. Many great scientists of the early twentieth century believed that, on the contrary, the universe was "always there." From 1917 until 1932, Einstein was among them. But on this count the Bible was right.

I am not advocating using Genesis as a source of information about the creation of the universe, but I am bringing up this point to show readers that science based on misconceptions will lead to wrong conclusions. Before we claim that we know exactly how the universe came into being, we had better check our science very carefully.

Interestingly, the Big Bang theory was not proposed by the astronomers who had discovered the expansion of the universe: Slipher, Humason, and Hubble. It came from a Belgian Catholic priest. In 1927, Georges Lemaître, an ordained priest who had studied mathematics at MIT, extrapolated the Hubble, Slipher, and Humason results backward in time to conclude that if the universe is expanding now, it must have been smaller and smaller as we go back into the past. In effect, he was able to use mathematics to rewind the movie of the progression of the universe to its very beginning and to demonstrate that it indeed had a beginning—as Scripture tells us.

Lemaître termed the germ of the universe the "primeval atom." He presented this theory of a Big Bang in mathematically rigorous papers that even today surprise researchers with their perceptiveness and accuracy. But Einstein, committed to his own equation that "forced" the universe to remain static, at first fought hard against the priest. *"Vos calculs sont corrects,"* he told Lemaître in French, *"mais votre physique est abominable."* ("Your calculations are correct, but your physics is abominable"). This was the first of several arguments that Einstein lost. The priest, following his mathematics, was right on the money.

This disagreement epitomizes a key problem of science: equations are only as good as the assumptions you make in formulating them. If the assumptions are false and do not reflect nature, the equations will lead to wrong conclusions even if they are set forth by the greatest scientific mind in the world.

We now know that the universe did have a beginning, the Big Bang. And through telescopic and satellite observations, as well as work done in particle accelerators such as the Large Hadron Collider, we believe we can understand how the universe progressed from a fraction of a second after its original "singularity" (Lemaître's "primeval atom": a place where the laws of physics as we know them do not hold) in the Big Bang to the present universe. But we do not—and perhaps cannot—know what caused the Big Bang or what, if anything, existed or happened before it.

When I interviewed the Nobel laureate physicist Steven Weinberg for a 2010 article about him in *Scientific American,* I asked: "How was the Big Bang caused, and what happened before it?" His answer was simple: "This we don't know, and have no way of knowing." This statement by one of the world's leading physicists and thinkers convinces me that science cannot disprove a "creator." If science cannot take us to the actual moment of creation and before it, then how can we argue against some preexisting essence and force that sent our universe on its way?

As we will see, some physicists have posited hypothetical models—since there are no data that emanate at the Big Bang or before it—that look at how the universe might have come about. But these models do not arise "out of nothing": there is always some preexisting substance, some kind of medium from which the universe emerges. (Often that medium is called *quantum foam*: a dense collection of bubbles of space and time in which space and time are highly interlinked and strongly mixed because of the effects of both general relativity and quantum mechanics.) There is really no logic in assuming that a universe comes out of nowhere; there must be something that precedes it.

Work in physics over the last century has led in the direction of a theory of the "unification of the forces." We identify four forces in nature: gravity, electromagnetism, and the weak and the strong forces that act inside nuclei. But theoretical advances (in particular, a theory called *supersymmetry*) have led physicists to the belief that the four forces of nature were probably once unified in a single force right after the Big Bang. This is called the *superforce*. It comes right out of the equations of physics as extrapolated back in time. But what was this superforce, a single, immensely powerful force of nature that governed our universe when it was very young? It is, in fact, something unknown and mysterious, and it is responsible for our being here. Some might call it God.

**BEFORE CONTINUING WITH** the science and its relation with religion, I feel compelled to counter certain points often raised by New Atheists about Einstein as a person. Einstein has been portrayed by several biographers as the consummate atheist, a "nonbeliever" and a "nonpracticing Jew." While Einstein did not generally adhere to the principles of any institutionalized religion, including his native Judaism, it's likewise not true that he was an atheist in the sense that New Atheists would like us to believe.

Einstein did attend synagogue services during his year in Prague, 1913, during the period of his highest scientific productivity. It appears he did believe in a kind of God: the entity that created the laws of nature, which Einstein viewed as his life's role to uncover.

Einstein spoke in terms of "God" all the time, famously saying, "Subtle is the Lord, but malicious he is not" (when a later-found-to-be spurious contention to relativity was brought to his attention), and "I want to know God's thoughts—the rest are details." These are hardly words that Richard Dawkins would be caught dead uttering. But Einstein made these statements—and many similar pronouncements about God—with clarity and conviction.

on, a letter Einstein sent to a little girl who had
to ask about his religious beliefs speaks volumes
about his attitude to God:

*Dear Phyllis,*

*I will attempt to reply to your question as simply as I can. Here is my answer:*

*Scientists believe that every occurrence, including the affairs of human beings, is due to the laws of nature. Therefore a scientist cannot be inclined to believe that the course of events can be influenced by prayer, that is, by a supernaturally manifested wish.*

*However, we must concede that our actual knowledge of these forces is imperfect, so that in the end the belief in the existence of a final, ultimate spirit rests on a kind of faith. Such belief remains widespread even with the current achievements in science.*

*But also, everyone who is seriously involved in the pursuit of science becomes convinced that some spirit is manifest in the laws of the universe, one that is vastly superior to that of man. In this way the pursuit of science leads to a religious feeling of a special sort, which is surely quite different from the religiosity of someone more naive.*

*With cordial greetings,*
*your A. Einstein*

In light of all this, to claim that Einstein was the most prominent scientific atheist of the modern age is a distortion of what Einstein was about. He viewed himself, perhaps allegorically, as an especially gifted human being on a life mission to uncover "God's thoughts," or at least God's laws of nature. So the poster boy (literally, given how ubiquitous his image is) for "scientific atheism" was not the man the atheistic crusaders would have you believe he was.

In *A Universe from Nothing*, Lawrence Krauss quotes Einstein: "What I want to know is whether God [sic] had any choice in the creation of the universe." The qualifier "[sic]" is usually inserted when pointing out a grammatical or spelling error in a quotation, or a gross misconception. Einstein doesn't need Lawrence Krauss to interpret his words for us.

Krauss then tries to "explain" what he believes Einstein actually meant:

> I have annotated this because Einstein's God was not the God of the Bible. For Einstein, the existence of order in the universe provided a sense of such profound wonder that he felt a spiritual attachment to it, which he labeled, motivated by Spinoza, with the moniker "God."

Given the many other references to God in his writings, on what basis does Krauss presume to interpret Einstein's words as if Einstein were illiterate and didn't know what he was saying?

But of course Krauss is imitating Richard Dawkins, who apparently felt so threatened by pronouncements about God by the greatest mind of the twentieth century that he was compelled to start his book by reinterpreting for us Einstein's words. In chapter 1 of *The God Delusion*, titled "A Deeply Religious Non-Believer," Dawkins argues that Einstein "didn't really mean it" when he talked about God. He quotes Einstein saying, "Science without religion is lame, religion without science is blind," but then quickly adds: "But Einstein also said, 'It was, of course, a lie what you read about my religious convictions, a lie which is being systematically repeated. I do not believe in a personal God and I have never denied this but have expressed it clearly.'" Dawkins goes on to say that people have cherry-picked Einstein's pronouncements about God. But of course Dawkins is guilty of the very same cherry-picking.

Einstein's relationship with God, or whatever he called God, is subtle and complex. In *Einstein: His Life and Times*, Philipp Frank, a gifted physicist and a close lifelong friend of Einstein, wrote, "The appointment as professor at Prague led Einstein to become a member of the Jewish religious community." Frank explains that the relationship was somewhat loose; Einstein also was embraced by the Jewish intellectual circle of prewar Prague: "At this time there was already a Jewish group that wanted to develop an independent intellectual life among the Jews. . . . This group was strongly influenced by the semi-mystical ideas of the Jewish philosopher Martin Buber. . . . Einstein was introduced to this group, met Franz Kafka, and became particularly friendly with Hugo Bergmann and Max Brod." Frank explains that this was a group that wanted to create Jewish cultural life not based on Orthodox Judaism, yet Jewish in its nature nonetheless.

When describing that same period in Einstein's life, the Prague year, Albrecht Fölsing writes in *Albert Einstein* about Einstein's relationship with religion. According to him, Einstein used the expression "poverty of ideas without faith," when referring to the Czechs and Germans of Prague, as compared with the Jewish intellectuals of that city. Further in his book, Fölsing says that Einstein was making use of "his re-assumed Jewish 'faith.'" He quotes Einstein saying, "I discovered that I was a Jew," and comments, "His experience in Prague may have struck a chord in him, for two years later—only five years after his arrival in Berlin—Einstein for the first time, and very decisively, avowed his Jewishness." Later, in Einstein's words, "This was a purely emotional reaction and was not based on the fact that substantial portions of our spiritual inheritance may have passed down to me."

We know that Einstein didn't believe in a "personal God," one who observes the actions of all people and actively intervenes in their lives. But the above instances and many others, as well as the pronouncements about God that Einstein made throughout his life in describing physics, make it clear that he did believe in a kind of superior power that "made" the laws of nature Einstein was bent on discovering. Thus Einstein should not be described as an atheist, and to say that "he didn't really mean" his repeated references to God is unwarranted.

# 6

## God and
## the Quantum

The word "revolution" would be too mild to describe the emergence of quantum theory, a new view of the processes of nature that take place on the tiny scale of atoms and elementary particles. Quantum theory was formulated in the 1920s by a group of mostly young physicists, key among them Erwin Schrödinger, Werner Heisenberg, Paul Dirac, Wolfgang Pauli, Niels Bohr, and Max Born.

Physics was turned on its head: causality, locality, and simultaneity came into question. In the quantum world, nothing resembles the world we know. In 1935, Erwin Schrödinger came up with the famous example of a cat alive and dead at the same time to illustrate the weirdness of the quantum world and the fact that in quantum mechanics, particles are in a *superposition* of states. A quantum particle can be both here *and* there at the same time—in the same way that the hypothetical cat is both dead *and* alive.

Schrödinger's thought experiment consists of placing a cat in a closed box. Inside the box there is a glass vial of cyanide attached to a mechanism that breaks the vial, releasing the cyanide and killing the cat, if an atom in a small amount of radioactive matter in the box disintegrates. Schrödinger's idea is that the atom's radioactive disintegration is a *quantum event*, hence controlled by the laws of quantum mechanics. The radioactive atom is in a state of *mixture* between disintegration and nondisintegration, which through the macroscopic mechanism of the vial gets transferred to Schrödinger's cat. Since we don't know whether the disintegration has taken place, the cat is in the *superposition* of two sates: alive and dead—until we open the box and thus collapse the wave function (a quantum entity is associated with a wave; a "collapse of the wave function" produces a definite state from a quantum superposition) and the cat is thrown into one of the two actual states: alive or dead.

In addition to the superposition of states, which exploits the wave nature of matter on the microscopic scale, there are many other phenomena that make quantum particles behave weirdly. Two or more particles can be so deeply intertwined—they are called "entangled"—that they act *as one* even if miles apart. This

*The quantum miracle in which a particle can*
*be in a superposition of two different states: as Schrödinger's cat,*
*which is both alive and dead simultaneously.*

idea actually originates with Einstein, who used it to attack the quantum picture of reality, which he did not like—even though he was one of the founders of quantum mechanics by discovering the photoelectric effect, in which light behaves as particles (it had previously been understood to behave as waves; today we know it to be *both*).

Einstein and two colleagues proposed in 1935 a "paradox" derived from quantum mechanics, now named after them as the Einstein-Podolsky-Rosen (EPR) paradox. Einstein attempted to use the EPR thought experiment to discredit the fledgling quantum theory (in this aspect, he failed; quantum theory remains valid). The paradox implied that if the wave nature of matter was taken seriously, then particles that had interacted in the past would continue to be entangled and whatever happened to one of them, as the wave function that governed both of them (even though they were now apart!) collapsed, the other particle would be forced to behave in the same way. For years, no one knew what to do about the EPR conundrum: if particles did act this way, it would shatter everything we believe about *locality*—because it would mean that something here can be instantaneously affected by something else at a faraway location.

The Northern Irish quantum theorist John Bell, working at the CERN laboratory, took Einstein seriously thirty years later, and in the 1960s published papers that presented what are called Bell's theorems, which could be used to detect entanglement in the real world. Eventually a series of experiments carried out in California by John Clauser and colleagues and in France by Alain Aspect at the University of Paris in Orsay have indeed verified the phenomenon of entanglement through many experiments: a particle in one location will act in concert with one that is across the room or across the universe, the "coordination" between them happening *instantaneously*—faster than a light signal that might travel from one particle to the other to convey any information.

As if all this strange behavior is not enough, in the quantum realm one cannot make out cause and effect: in the quantum world you could not tell if a match thrown on the ground caused a forest fire or if the forest fire caused the match to light. To get a handle on the theory, researchers found they needed to resort to probabilities.

But the "scientific atheists" of our day use these same strange probabilistic rules of quantum mechanics to argue that God does not exist—that these laws somehow replace God. And that since we have these quantum laws—which we have highly incomplete understanding of—there is no need for a "creator." According to Lawrence Krauss, "we all, literally, emerged from quantum nothingness." But quantum rules do not at all imply that a universe must appear out of the void.

Besides the fact that we do not fully understand quantum theory, there is no well-defined boundary: we have not identified a point, a scale of measurement, at which things stop behaving according to our everyday life rules and start acting according to the bizarre quantum laws.

A good scientific theory is defined as one that allows us to make valid, verified predictions about future observations. But quantum theory is so weird that its predictions are usually understood only as *probabilities*. Quantum mechanics relies on the understanding that a particle is also a wave. The wave can lead to a probability distribution for various actual outcomes of an experiment. According to the standard, or Copenhagen, interpretation (the German physicist Werner Heisenberg developed this interpretation while working in Copenhagen with the Danish quantum pioneer Niels Bohr, who was his mentor), we can only predict the *probability* of a given outcome of an experiment—not what will actually happen. According to Heisenberg and Bohr, the wave quality of a particle *collapses* when we make a measurement of the particle. The actual measurement is a realization from a probability distribution (which is the square of the amplitude of the wave function).

An alternative, though less plausible, approach is the "many-worlds" interpretation of quantum mechanics, proposed by Hugh Everett III, which claims something even more bizarre than probabilities: that what does not happen here in any given experiment actually takes place in *another world*. We conduct an experiment and obtain an outcome out of the many possibilities inherent in the wave function of any particle. Since other possible outcomes *could also have* occurred, according to Everett, they *do occur*—in other worlds.

But if a theory cannot predict actual outcomes it cannot impart to us perfect knowledge. Thus, it is highly suspect to employ quantum physics to attack the existence of God. This is a strong argument against the New Atheists who claim that quantum mechanics "tells us" a universe will appear from the void, a claim I will return to.

One of the defining moments in my career as a mathematician and scientist took place when, in the fall of 1972, I met one of the fathers of quantum theory, Heisenberg, who came that year to visit the physics department at the University of California at Berkeley while I was a physics student there. Heisenberg gave us a brilliant talk about his discovery of the famous uncertainty principle that governs quantum behavior.

Heisenberg's uncertainty principle says that the product of the uncertainties in a particle's momentum and position must be at least as large as a constant (related to Planck's constant, a number specified by the German physicist Max Planck). If we measure the particle's position, we disturb it by the mere act of measuring, and therefore if we *then* measure the momentum, we get a different answer from the one we would have gotten if we had measured the momentum *first*; measuring the momentum of the particle first would disturb it, and its position, measured next, would be different from what it would have been if it had been measured first.

The uncertainty principle governs everything in the quantum world: *levels of variables are not known with certainty*. In its most commonly used form, Heisenberg's uncertainty principle is extended beyond a particle's location and momentum and applies to the two most important concepts in physical science: *energy* and *time*.

The uncertainty principle says that on the micro level of atoms and molecules and smaller particles, you can't *know* things with precision, only to within a statistical approximation. If you know the energy precisely, then you don't know with certainty the time measurement associated with that level of energy; and if you know the time precisely, then you don't know the energy with certainty.

Quantum theory allows us to make probabilistic or statistical predictions, although it also predicts the values of constants of nature. Quantum mechanics can tell us with unprecedented precision the probabilistic outcomes of an experiment. If, in an experiment, quantum theory tells us that there is a 50 percent probability that a particle will be measured to have its spin "up" and a 50 percent probability that it will show a spin "down," then if we perform the experiment using a million particle measurements, we can be sure that we would indeed find very close to half a million "up" measurements and half a million "down" measurements.

The theory has also been extremely successful in predicting the exact values of energy levels of the hydrogen atom (including something called the Lamb shift, which has been explained using interactions of an electron with virtual particles in the "vacuum").

The equation derived by Erwin Schrödinger in 1925, published on January 1, 1926, and known as the Schrödinger equation, uses the wave properties of matter, discovered some years earlier by Louis de Broglie. This "wave equation" is a differential equation that governs the behavior of quantum particles seen as waves. The uncertainty of the quantum world appears here as well as in Heisenberg's work because waves undulate and their undulation can be interpreted (when squared) as a probability distribution for the properties of the particles that possess the wave function—meaning all small particles governed by the laws of quantum mechanics.

We know that waves are additive. You can add two waves constructively—think of two waves in the ocean where one outruns the earlier one, producing a bigger wave—or destructively, where the trough of one wave coincides with the peak of another, thus making the sum of the two waves flat: the wave amplitudes cancel each other out.

It is this wave nature of particles that makes the quantum

world what it is and creates its weird behavior: it allows for the *superposition* of states. (The cat in the thought experiment is in a superposition of being dead and alive.)

*The wave functions in quantum mechanics can be added and subtracted, just like two ocean waves that come together to make one big wave (or one smaller wave when they cancel each other out).*

Richard Feynman extended the superposition idea to develop a theory in which a particle goes from one location to another using "all possible paths." Thus, to go from point A to point B a particle may take not only the direct straight-line route between them but also follow "paths that visit the restaurant that serves that great curried shrimp, and then circle Jupiter a few times before heading home," as Hawking and Mlodinow describe it in their book *The Grand Design*. Each path from A to B gets assigned a probability, and the probabilities lead to the most likely path being most dominant in the final calculation.

But according to this outlandish theory—whose predictions, however, have excellent verification in experiments, so it "works"—there is never a definite "history" to any process: the particle in this example took *all* paths from A to B, with different probabilities. Hawking and Mlodinow use this idea of Feynman's to conclude that the universe had *no definite history*.

What they mean is that in the same way that a particle goes from one location to another taking all possible paths—whatever that may actually mean—so does the entire universe. The reasoning is that during the Big Bang or a fraction of a second afterward, while it was tiny and very compact—perhaps the size of an atom or less—the universe had to follow the quantum rules. If it took "all possible paths" from beginning to getting to the size of, say, a grain of sand, a small macro object, then the universe has had no definite history. Hawking and Mlodinow say:

> Quantum physics tells us that no matter how thorough our observation of the present, the (unobserved) past, like the future, is indefinite and exists only as a spectrum of possibilities. The universe, according to quantum physics, has no single past, or history.

No single past or history? This is a stunning theoretical assertion. It shows that in the otherworldly realm of the quantum such bizarre things can happen that they shatter our concept of a "history." And since our entire universe is believed to have once been the size of a quantum particle, strange things can be hypothesized about its past. Richard Feynman's "all paths" method derives from the oldest quantum experiment ever undertaken, called the double-slit experiment, which was carried out in 1803 by the British medical doctor Thomas Young, who was interested in many things and even deciphered Egyptian hieroglyphics. In the experiment Young performed, light was passed through two slits cut into a screen and projected onto another screen, where an interference pattern became evident. This proved that light was a wave, like waves in water (Young also demonstrated interference of water waves, for comparison). What happens here, however, becomes a big mystery in quantum physics, and Feynman called it *the* mystery of quantum mechanics.

As Alain Aspect explained it to me when I visited him in his lab in Paris, "With modern light sources, it is possible to control the emission of light so that the lamp releases one photon at a time." When a single photon is released and encounters the two slits, the interference pattern *still* forms. This means that, somehow, the photon goes through *both* slits, and then interferes with *itself.* This is the simplest example of the superposition of states: the photon is in two places at once going through slit 1 *and* going through slit 2. Feynman was right to view this phenomenon as an embodiment of all the mysteries of quantum mechanics.

In a fanciful description of how Feynman might have arrived at his all-paths approach to quantum mechanics, the theoretical physicist A. Zee wrote in his book *Quantum Field Theory in a Nutshell*:

> Long ago, in a quantum mechanics class, the professor droned on and on about the double-slit experiment, giving the standard treatment. . . . Suddenly, a very bright student, let us call him Feynman, asked, "Professor, what if we drill a third hole in the screen?" The professor replied, "Clearly, the amplitude for the particle to be detected at the point *O* is now given by the sum of three amplitudes. . . ." The professor was just about ready to continue when Feynman interjected again, "What if I drill a fourth and a fifth hole in the screen?" Now the professor is visibly losing his patience: "All right, wise guy, I think it is obvious to the whole class that we just sum over all the holes.". . . But Feynman persisted, "What if we now add another screen with some holes drilled in it?" The professor was really losing his patience. . . . Feynman continued to pester, "What if I put a third screen, a fourth screen, eh? What if I put a screen and drill an infinite number of holes in it so that the screen is no longer there?" The professor sighed, "Let's move on, there is a lot of material to cover in this course."

Zee then agrees with Feynman that indeed the two-slit experiment can conceptually be extended to an infinity of holes.

He adds: "Surely you see what that wise guy Feynman was driving at. I especially enjoy his observation that if you put in a screen and drill an infinite number of holes in it, then the screen is not really there. Very Zen!" What Feynman showed was that even if there were empty space between the source and the detector—no barrier with slits at all—the particle should still take "all possible paths" from source to destination. Calculations made using this theory verify experimental findings, but it is important to note that very "exotic" paths from source to destination are given extremely small probabilities. Still, this is a very strange way of viewing reality, and perhaps reflects the fact that in quantum mechanics we *have no real understanding* of reality.

The renowned French philosopher and quantum theorist Bernard d'Espagnat concludes that because quantum theory is so incomprehensible, it is not a "real" theory. He calls it a *veiled reality theory*. What we see in the quantum world is thus a "veiled" version of what is really going on "inside" the black box of reality at the micro level. D'Espagnat explains:

> The Veiled Reality conception . . . merely involves the conjecture that our great mathematical laws are highly distorted reflections—or traces impossible to decipher with certainty—of the great structures of "the Real."

Because of the immense conceptual difficulties in obtaining any deep, fundamental understanding of the reality behind quantum observations and calculations, d'Espagnat tried to solve the problem using an interesting analogy, paraphrasing writings by the French physicist Hervé Zwirn:

> According to all appearances, our abilities at conceptualizing exceed those of dogs, monkeys, and other animals. And he [Zwirn] wondered whether or not it is possible to envisage a conceptualizing ability exceeding ours in the same way as ours exceeds the ones of dogs and monkeys. He pointed out that to bluntly answer "no" would be highly presumptuous. . . . We are thus left with the other alternative, which is to grant that, after all, there is no absurdity in evoking the idea of a "something" that we cannot conceptualize.

According to d'Espagnat, even "cosmic time" doesn't have a meaning as being absolute in any sense, and he posits that the religious idea of immortality could be interpreted by the physicist as follows:

> The term "immortality". . . has become somewhat puzzling since it seems to implicitly postulate an absolute time conceptually anterior to the human mind. The question then is, shouldn't this term rather be understood to refer, in the picturesque style religions are forced to use, to another notion that also be-

longs to their realm, namely that of "eternity," in the
sense of *escaping* from time? For the same reasons we
should (even more daringly!) examine whether or not
also the notion "creation," the "creative act," might,
by means of some refocusing on the one Being, be
made independent of time; at least of the human ex-
perienced time, the time of empirical reality.

D'Espagnat, having considered the same quantum weirdness of the universe as had others, has come to a very different
conclusion about what we as human beings can understand and
conceptualize. He is inclined to believe in creation, as well as
in eternity. Perhaps thinking about the mysteries of quantum
mechanics has brought d'Espagnat to these ideas that seem to
wed faith with science.

The strangeness of quantum theory has forced not only
d'Espagnat but several other quantum pioneers, such as John
Bell, to conclude that the theory cannot tell us what *is* but rather
only show us some shadow of the reality we cannot reach. Bell
was the Northern Irish theoretical physicist who explained and
elucidated the concept of entanglement—Einstein's "spooky
action at a distance," which he thought was so weird that it
could never happen—and yet Bell proved that it could.

In his book *Speakable and Unspeakable in Quantum Mechanics*
(1993), Bell speculates about the deep truths of quantum mechanics and their relationship with reality as we understand it.
Bell looks at various categories of entities involved in quantum

theory: observables (things we observe), the experimental apparatus we use in quantum experiments, the variables in each situation—ones we control and ones we do not—and so on. Bell then proposes the concept of a "beable." A beable is a real element of the theory, while others are not. Bell calls things that are physical "beables" (things that are real, rather than mere mathematical artifacts). According to Bell, fields (such as the magnetic field of the Earth) are beables—they have a reality to them—but potentials (such as the electromagnetic potential, used in calculations in physics) are not. The potential "is not really supposed to *be* there. It is just a mathematical convenience." He continues:

> One of the apparent non-localities of quantum mechanics is the instantaneous, over all space, "collapse of the wave function" on "measurement." But this does not bother us if we do not grant beable status to the wave function. We can regard it simply as a convenient but inessential mathematical device for formulating correlations between experimental procedures and experimental results, i.e., between one set of beables and another.

Both Bell and d'Espagnat draw a line between what we can understand when dealing with the quantum world and what we cannot, and believe that the underlying "reality" of quantum mechanics is well outside our comprehension. Quantum theory may well be one of those things that could only be truly understood by a super-human being: something that, compared to us, would be like us compared with dogs or monkeys.

We simply do not understand what goes on in the world of the very small. Gerard 't Hooft, the Dutch Nobel laureate particle physicist who solved one of the most complicated theoretical conundrums in quantum field theory, recently told me that his new research is going back to basics in trying to gain us "a better understanding of quantum mechanics." Once again, we still don't understand at all what truly happens in the world of the very small—all we have may be shadows on the wall, cast by a mysterious "veiled reality."

So to claim that quantum mechanics somehow "tells us" that a universe must come out of nothing without the need of some kind of creation, as Krauss does, seems wholly unjustified. It arises, in part, from a misinterpretation of how particles are believed to emerge in the universe. We will discuss this idea next.

# 7

## The "Universe from Nothing" Deception

I n 1928, Paul Dirac wedded quantum mechanics with Einstein's special theory of relativity, creating something called relativistic quantum field theory. Dirac's equation led him to infer the existence of antimatter—which, just a few years later, was confirmed experimentally. This idea shows that particles can actually be *created* in pairs out of sheer energy. Empty space has been shown experimentally to be permeated with energy, which turns into pairs of "virtual particle pairs" all the time. There is no such thing as purely empty space: because fields of various kinds exist throughout any section of space, space is teeming with pairs of electrons and anti-electrons (called positrons), for example, which pop in and out of existence as energy turns into pairs of particles that then annihilate each other to form gamma rays, a form of energy.

Dirac's prediction of the existence of antimatter, and its later experimental discovery, were a monumental advance in physics. The existence of antimatter is understood within the context of Einstein's famous formula, $E = mc^2$, which equates the concepts of mass and energy. Dirac's theory of antimatter provided the actual mechanism for "creation."

How? Einstein taught us that mass is energy and energy is mass. According to relativity theory, the two are really one and the same thing. But until Dirac, we didn't know how to *change* one essence into the other: how to change mass into energy, and energy into mass. If, as Einstein says, the two are one and the same thing, they should be interchangeable. Dirac's famous equation, which incorporates the stipulations of Einstein's special theory of relativity, provides the actual *mechanism* by which mass can change into energy and energy can change into mass. This mechanism is called pair production. The term means that from a given amount of energy, nature can *create* matched pairs of particles.

The universe has beautiful symmetries embedded within its very framework, and the pair of particles that can be produced through Dirac's mechanism from a small amount of energy is a perfectly matched pair of elements: they are mirror images in every way, one particle made of matter and the other of antimatter. Since the photon (a quantum of energy) that leads to the creation of the pair has no electric charge, one of the particles created will have a positive charge and the other will have an equal amount of negative charge. The two particles are like yin and yang: one complements the other, and their total charge, together, is zero—the original charge of the photon from which they were spawned.

So, once you have some energy available to you, then using quantum rules and Dirac's equation you could actually create pairs of particles and antiparticles, and hence (if you make enough of them) a universe seemingly "out of nothing." Of course it's not out of nothing—you require the energy to make the particles via Dirac's mechanism. The assumption is that the energy comes from some "quantum fluctuation"—the energy is produced through a quantum undulation in something. It cannot be an undulation in a complete emptiness.

Lawrence Krauss has misused the idea of "empty space" to argue that the universe itself came out of sheer "emptiness." But we know that the space in which pairs of particles can form is never empty, it is not a "nothing"—it *always* contains energy, and it always becomes permeated by lines of force representing fields (electromagnetic, gravitational, and other); and it is the energy supplied by these fields that leads to the creation of pairs of particles. The creation of such particles is therefore never "out of nothing"—it is out of a preexisting space that is filled with energy. That space, that energy, and the fields that permeate it all have to come from somewhere. But there are many problems even here that have not been addressed by this theory. First, what happens to the antimatter? Where does it all go? Why doesn't it annihilate the matter, again creating energy? And the quantum fluctuation in energy, which creates the pairs of particles has to be a fluctuation in *something*. What is that *something* in which the fluctuation occurs? It has to be something that preexists.

Lawrence Krauss's book *A Universe from Nothing* claims that the universe came about out of sheer nothingness, with nothing preexisting. For proof, Krauss relies on a paper written by the physicist Alexander Vilenkin (which he does not even reference correctly), "Quantum Origin of the Universe," presenting a discussion of theoretical physics, including both general relativity and quantum mechanics, to attempt to trace the very origin of our universe.

In the last page of his article, Vilenkin notes the following:

> Most of the problems discussed in this paper belong to "metaphysical cosmology," which is the branch of cosmology totally decoupled from observations. This does not mean, however, that such problems do not allow rational analysis: the ideas can be tested by overall consistency of our picture of the universe.

And the picture of the universe that Vilenkin uses consists of what we know about the universe through theory and observation. The starting point of his cosmological model is, indeed, something that is "earlier" in a sense than the existence of some piece of space-time that then fills up with fields and then particles and consequently grows to become the universe we know. Vilenkin's "nothingness" is indeed more "nothingy" than anything that had been contemplated before. But, careful scientist that he is, Vilenkin is still cautious about his definition of nothingness, usually placing the term inside quotation marks.

Vilenkin's nothingness is not an existing space-time. It is a single point, with no extent. But that point is still *embedded in a preexisting medium*: the *quantum foam* that existed before the creation of our universe. (Quantum foam is a highly turbulent, condensed medium in which space and time are highly curved and in which quantum effects, and the effects of general relativity, are very strong.) In that sense, the universe did not come about from *absolute and complete nothingness*, something like the mathematician's empty set. The "nothingness" of Vilenkin's model is the absence of a classical space-time. He writes, "I shall discuss a model in which the universe is created by quantum tunneling from 'nothing,' where by 'nothing' I mean a space with no classical space time."

Using the quantum-mechanical process called quantum tunneling, through which a quantum system (usually a small particle, but here that small particle is the extentless point that would become the universe) "tunnels" through a classical boundary. It can do this because, as we recall, in quantum mechanics a particle is also a wave, and therefore has a wave function associated with it. It can happen that the wavelength of the wave function extends beyond the location of some physical barrier such as a thin piece of metal. Now, since the wave function, when squared, results in a probability function, it may happen that the probability function "stretches" beyond the boundary. This means that the particle has a given nonzero probability of *actually existing beyond the boundary*. And, since in physics anything that *can* happen (meaning has a nonzero probability of occurring) eventually will happen, the particle can "tunnel through" to the other side of the barrier.

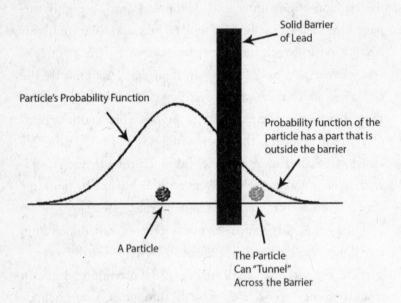

*A quantum particle can "tunnel" its way through a solid barrier because its wave and probability functions may extend beyond something solid.*

Applying this idea to the point representing the unexpanded universe, Vilenkin argues that its wave function can allow it to extend beyond its "nothingness" state—and hence it eventually does. And the rest, as they say, is history: the Big Bang!

But Vilenkin's nothingness is not absolute nothingness. He requires many things for his universe to come about: quantum foam, Einstein's gravitational field, the Higgs field, quantum tunneling, and other physical entities and laws. Therefore, to claim, as Krauss does in his book, that thus a universe can be created out of sheer nothingness is deceptive.

Not beholden to the Dawkins doctrine of disproving the existence of God, Vilenkin himself actually goes somewhat in the other direction. In the introduction to his paper, he writes: "The idea that the universe was created from nothing is at least as old as the Old Testament." Then he quotes St. Augustine (*Confessions*): "What was God doing before He made heaven and earth? If He was at rest and doing nothing, why did He not continue to do nothing for ever after, as for ever before?" The answer, according to Vilenkin, is quantum tunneling from "nothingness." Space and time are created in the Big Bang that results from the quantum fluctuation in the foam:

QUANTUM FOAM

QUANTUM FLUCTUATION (VIA QUANTUM TUNNELING)

UNIVERSE

Once the quantum effects take place, according to this model, Einstein's general relativity comes into play, and we find the universe we have today, dominated by gravity and the other three forces of nature.

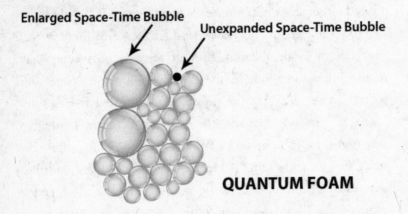

**QUANTUM FOAM**

*Bubbles in the dense space–time mixture called quantum foam.*

So Vilenkin's universe does not really start from "nothingness" at all. Vilenkin's universe begins with *an actual, existing bubble within the preexisting quantum foam*. This is what he writes: "It describes a bubble which contracts at (time) t<0, then bounces at a minimum size $R_0$, and expands at t>0. In the actual history of the bubble the t<0 part is absent: the bubble tunnels quantum-mechanically from $R = 0$ to $R = R_0$, and then evolves. . . ." What Vilenkin describes here is a situation where in an actual "bubble" in the preexisting quantum foam, some-

how the internal energy (viewed as a kind of pressure) is lower than outside the bubble. So, just as would happen to a chewing gum bubble that lost pressure, the quantum bubble contracts as the outside pressure overcomes the internal pressure, until the bubble shrinks to zero radius, and then—through the miracle of quantum tunneling—it bounces back to larger volume and from there expands in a Big Bang. The trick here is that the contraction to zero size (the "nothingness") occurs at a time that is *defined as negative* (t<0), and hence we can never "observe" it, since we can only "know" things that happen after the concept of time exists for us, i.e., after the Big Bang—which is the creation of space and of time.

**VILENKIN'S IDEA ACTUALLY** originates, he tells us in his paper, in an earlier, revolutionary concept proposed by the physicist Edward P. Tryon. MIT physicist Alan Guth also credits Tryon as the originator of the "universe out of nothing" idea. Tryon's 1973 paper "Is the Universe a Vacuum Fluctuation?" proposes the idea that our universe may have a sum of zero for all its *quantum numbers*—quantities such as electric charge that physicists believe are "conserved" in nature, meaning that they cannot be created or destroyed (remember that in pair production from a neutral photon, one particle must be positively charged and the other have equal but negative charge—this is the manifestation of a conservation law).

Energy is the most important quantity believed to be conserved in nature. Energy cannot be created or destroyed—it can

only *change* from one form to another, as when energy turns into mass and mass turns into energy according to Einstein's famous equation. Tryon argued—perhaps for the first time—that the total energy of the universe was zero. Because energy is believed to be conserved in the universe, it cannot be "created." This means that the energy we see in the universe around us either *always existed* or that God created it at some point. (Presumably, God can break any human-proposed "conservation law" in physics.) But, if the total energy of the universe is identically zero, as Tryon thinks, then no "creation" is necessary. This is because zero is zero and will always remain zero. And in that case, the zero energy simply changed form to *positive plus negative energies*. Then ultimately, according to Tryon, energies turned into a universe of matter coexisting with a universe of antimatter.

But today there's still no answer to the question of where all the antimatter is that theoretically came into being along with regular matter in the Big Bang. Most physicists believe that the balance of matter and antimatter is *not* conserved. A form of theoretical imbalance between matter and antimatter is called a CP violation, where "C" stands for *charge conjugation* (reversal of positive and negative electric charges) and "P" stands for *parity* (essentially reflecting through a mirror, which exchanges left with right).

Until the mid-1950s, physicists believed that parity was a conserved quantity in nature. This means that if you look at a physical system, then hold a mirror to it and look at it through the mirror, you will observe exactly the same physics, alas in

reversed direction: a particle moving to the right will be progressing to the left. This kind of idea was attractive, for why would reflection in a mirror change the essentials of the physical processes? But in fact, it does. The idea of the violation of the parity conservation law occurred to two American theoretical physicists of Chinese origin, C. N. Yang and T. D. Lee, in the 1950s while they were having dinner at a Chinese restaurant in New York.

Together, they came up with the idea that perhaps the weak nuclear force acting inside the nucleus of an atom did not "respect" parity. They suggested to another Chinese American physicist, known affectionately to physicists as "Madame Wu" (Chien-Shiung Wu), at Columbia University, that she carry out an experiment to test their theory. She confirmed their strange prediction: parity is not a conserved quantity, at least with respect to the weak nuclear force. Later, it was discovered that when we put parity together with charge conjugation, thus making a "quantity" called CP, that quantity, too, is not conserved in nature. Thus, there is a difference albeit perhaps a very small one—between matter and antimatter, since changing from matter to antimatter means "looking in the mirror and reversing all electric charges."

Such "CP violation," as it is known, may or may not be strong enough to have resulted in a universe of only matter and no antimatter. The question is still open, and it is one of the most important ones in all of physics and cosmology. On the other hand, there may be an entire universe made of pure antimatter

that exists somewhere in space, perhaps symmetrically opposite to our own universe, if matter and antimatter were indeed created in equal amounts from the burst of sheer energy that was the Big Bang.

Tryon's other arguments rely on the universe being "closed," having zero total energy. But because of the 1998 astronomical discovery of the accelerating expansion of the universe, and evidence from the microwave background radiation obtained from satellite observations, we know today that the universe is *not* "closed." Rather, our universe is very close to being perfectly "flat." This means that, overall, space has the geometry of Euclid—characterized by straight lines, rather than having any intrinsic overall curvature. So Tryon's argument completely fails given what we have learned about the universe in the decades since his paper was published.

Equally, the fact that the universe is believed to contain an immense amount of "dark energy," whose exact nature and quantity are still unknown, makes any argument about "zero total energy" for our universe a moot point.

A "universe out of nothing" in the sense of Lawrence M. Krauss is still a figment of the imagination and has no basis in any objective reality. It does not arise out of the theoretical results of Vilenkin or Tryon, since their universes do not emerge from absolute "nothingness": Vilenkin requires the preexisting quantum foam and a "hidden history" of the pointlike baby universe at a time we cannot observe ("negative time"); and Tryon needs a closed universe, which we now know ours is not, and he

doesn't address the disappearance of antimatter, on which his theory also relies.

**BUT OUR UNIVERSE** may well have been derived from a quantum fluctuation in the preexisting quantum foam that likely preceded the Big Bang. The required quantum foam, with no classical space and time, in Vilenkin's model, implies that the quantum fluctuation that produced our universe might have happened more than once—following the philosophy that anything that *can* happen eventually *will* take place. Thus there may be, according to some cosmologists, a *multiverse*: a multitude of universes. Using Augustine's language, if God stopped doing nothing once, he might as well have stopped doing nothing again and again.

# 8

## And on the Eighth Day, God Created the Multiverse

On March 2, 2011, I took part in the most unpleasant public debate in my life. It was with a physicist I had once considered a friend, or at least a friendly acquaintance. Brian Greene had just written a book called *The Hidden Reality: Parallel Universes and the Deep Laws of the Cosmos* and was touring the United States to speak about it.

I was asked by the Boston Museum of Science to interview Greene about his book in a televised event aired on C-SPAN. Having known Greene for a number of years, I gladly agreed. But the event turned sour. Almost every time I asked him for elaboration of his thesis that there are other universes, or for any experimental proof of what he was saying, Greene would evade the question.

Capable scientist that he is, he presented no convincing

proof that other universes do exist out there. All he would say was something like: "The math tells us so, and I believe in the math." Mathematics, however, does not tell us anything about other universes—*real* universes, that is. All the routes to the "multiverse"—the collection of universes that may or may not exist out there—are wholly hypothetical. The debate was frustrating mainly because of the fact that we were dealing with topics that are doubtful and far outside anything that is knowable to us as scientists in any objective way. We might as well have been debating how many angels can dance on the head of a pin.

Atheists have latched on to the idea of a multiverse, assuming that if many universes exist, then the creation of one universe seems less spectacular and thus perhaps doesn't require a divine act. They welcomed Greene's book.

In his book, Greene had put together four distinct theories that he argued implied that our universe is only one of many—perhaps infinitely many—entire universes: some like our own, and others not.

One such theory was Alan Guth's inflationary universe hypothesis, which says that at its inception, the universe went through a period of very rapid expansion (called inflation), which later slowed down significantly. That theory has been extended by Andrei Linde and Alexander Vilenkin to *eternal inflation*. Linde and Vilenkin believe that, because of quantum considerations, the inflation process that created our universe is forever ongoing.

According to these theorists, the inflation process simply "moves elsewhere" in the larger universe, where it continues to inflate other segments—ones that are inaccessible to us because of the immense distances created by the rapid expansion of space. And as these small pieces of the universe grow large extremely fast, they become even more distant from us, and now can be seen as independent "universes" because they are so far away and unreachable to us.

The question is: What is the value of such an assertion? It explains how parts of our universe may go their own way as inflation takes over in them once it has stopped inflating our own part of the universe. All well and good. But this is not the milieu of a true "multiverse"—it's just a theoretical setting in which there are far-off segments of the universe of which we are a part. And we are not really sure that this theory is true. Physicists just don't know how to theoretically "stop" inflation once it has started, and since we know that our own part of the universe is no longer inflating (it is expanding at a more moderate pace than that of inflation), the assumption is that inflation moves to some other location. Also, since we can never observe such distant parts of our own universe, or obtain any information from them, what's the use of such a model?

**HUGH EVERETT'S MANY-WORLDS** interpretation of quantum mechanics is another route to the multiverse, and one that Brian Greene supports. Many-worlds is even more unlikely than the eternal inflation theory. It tells us that because we don't have a good theoretical way of "collapsing the wave function" of quantum mechanics (meaning making the fuzzy entities of quantum reality become definite), every possibility—every potential outcome of an experiment we may carry out—that has *not* taken place here, happens in some "other universe."

Just where these universes may be, we don't know. And there are so *many* of them: every possible outcome of every quantum event in our world goes into another universe! These quantum events happen all the time, everywhere: every instance when a photon is released in a lightbulb because an electron drops from a higher to a lower energy level is a quantum event. And every chemical reaction involves quantum events. There are an inconceivably large number of such occurrences happening all the time.

If you drive and decide to turn left at an intersection, then, according to many-worlds, there is a universe that is like ours but in which you turned right. There are universes where Hitler won World War II and in which Nazis rule the world; and there are universes where 9/11 didn't happen and the World Trade Center is still standing. There is no experimental support for embracing such a weird theory, and it generally has few adherents.

**STRING THEORY IS** yet another area of physics that has led Greene and others to propose the existence of many universes. What is string theory? It's an approach to physics that has been developed over the past four decades, in which the basic elements of nature are tiny vibrating strings. String theory was first proposed by the Italian physicist Gabriele Veneziano while he was spending a year as a young researcher in the 1960s at the Weizmann Institute in Israel.

"I was looking at the equations governing the motions of particles one day," he told me when I interviewed him in Genoa, Italy, in 2005, "when I noticed that they resembled those of strings, just like the strings of a violin." He checked his work and, indeed, there was a relationship between string vibrations and the motions of tiny particles. Thus string theory was born. But from there it went very, very far. Mathematical physicists, prominent among them Edward Witten of the Institute for Advanced Study at Princeton, took over and introduced into string theory mathematical methods so advanced, so powerful, and so novel that string theory is now almost considered a branch of pure mathematics. In fact, these innovations won Witten the Fields Medal, the top prize in mathematics.

String theory has what mathematical physicists view as elegance, and many have found it an exciting area to work in. But this theory has not produced much that can be verified by experimentation or even be replicated through other approaches. "Their only success is the entropy of a black hole," Roger Penrose told me when I interviewed him, referring to the fact that string theory has reproduced the result of a theoretical computation of a physical measure of a black hole using other techniques. And at present there is no way to even run an experiment to confirm any of the predictions of this highly theoretical field of study.

In string theory, the universe is seen to inhabit a space that has more than the usual four dimensions of space and time because the equations that govern the behavior of these strings make sense, mathematically, only in a larger context of ten or eleven dimensions. Some scientists have taken this *theoretical* requirement of string theory to mean that the *actual* physical universe in which we live must have an extra six or seven latent dimensions. String theorists such as Greene call these "curled-up dimensions," viewing them as hidden within the three dimensions of space and one of time that we are aware of.

The big question is whether the fact that certain equations use more than four dimensions indeed implies that the actual universe they describe must also have extra dimensions. To paraphrase the quantum theorist John Bell, are these extra dimensions "beables"—are they *real*—or are they simply due to a mathematical convenience?

Since string theory has not yet provided any verifiable predictions and is not likely to offer them any time soon, the actual—rather than merely mathematical—existence of the extra dimensions remains a mystery. Are these dimensions only a mathematical oddity, a mathematical requirement of the theory, or do they tell us something real about our universe?

Greene and his colleagues have used the additional dimensions of string theory to argue that other universes may "hide" somewhere within these extra dimensions. But again, given that none of string theory's predictions can be verified within our experimental capabilities, it's dubious to hypothesize that there are other "hidden" universes lurking in the background.

**A FOURTH LINE** of reasoning for the possible existence of other universes that Greene entertains is the *anthropic principle*. This principle, which I will discuss in detail soon, has led some physicists to propose that, while our universe may well have been a onetime event, there are other universes that we can't see because we can only observe a place that is conducive to our existence, and cannot observe places whose environments are hostile to life. The idea is that there are many things that we do not understand about our universe: its parameters seem to be too "fine tuned" to have arisen by chance, and since all their values are set at the exact levels required for our existence, there "must be" other places—other universes—where these parameters are *not* right.

Basically, to avoid the necessity of "creation"—which *begs* to be acknowledged as a possibility in a universe that seems almost "too perfect" in its ability to support life—these physicists take the following view: if we are here, and the parameters need to be perfectly chosen for us to be here, then surely there must be infinitely many other places where the parameters are wrong. We are here because we can only live where the conditions are right for our existence.

The problem with this route to the multiverse is that it does not even provide a *mechanism* for the creation of the other universes we don't see. For all their faults, at least the eternal inflation theory, string theory, and many-worlds offer this much. The anthropic theory is the weakest route to the multiverse.

**JUST BECAUSE ABSTRACT** equations may require more dimensions than we observe does not mean that such dimensions are a reality. Just because we don't know how to "stop" inflation doesn't mean that it creates other universes. And just because we understand so little about the wave function of quantum mechanics doesn't mean that a wave can live on in other worlds.

New Atheists have embraced the multiverse idea—speculative as it may be—because it appears to do away with a creator. To them, the laws of physics and mathematics lead to the creation of a universe out of nothing; and if this can happen once, then it can happen again and again, hence the possibility of even an infinite number of universes. And if infinitely many universes exist, ours is of infinitesimal importance and so perhaps would have required no divine powers to create. But an alternate argument says that if many, perhaps even infinitely many, universes exist, then whatever created them must be much more powerful than anything we've ever contemplated before. In any case, we can only observe one universe.

The worst feature of the multiverse theory is that it is *nonparsimonious*. It is a model that—like Ptolemy's ancient theory about the solar system with its cycles and epicycles, so decisively smashed by Copernicus—has too many free parameters. In fact, an infinite multiverse has *infinitely many* parameters. There must be parameters to describe each of these infinitely many other universes that are supposed to somehow exist out there. The infinite multiverse would never pass any Einstein test for elegance or simplicity. And nature tends to be captured best by models that are both.

But even the non-mathematical Dawkins seems to be captivated by the promise of the multiverse as a way of avoiding God. Here is what he writes:

> It is tempting to think (and many have succumbed) that to postulate a plethora of universes is a profligate luxury which should not be allowed. If we are going to permit the extravagance of a multiverse, so the argument runs, we might as well be hung for a sheep as a lamb and allow a God. Aren't they both equally unparsimonious ad hoc hypotheses, and equally unsatisfactory? People who think that have not had their consciousness raised by natural selection.

Dawkins grossly underestimates just how "profligate" the "plethora of universes" idea really is. And why he thinks that the physical multiverse could have anything in common with biological "natural selection," and how one could "raise consciousness" by contemplating natural selection in order to understand an infinite set of universes, is really anyone's guess.

The main problem with the multiverse is that there is absolutely no way in which we can validate such theories through experimentation or through any data obtained or derived from the real world. And an infinite multiverse invokes mathematical principles that are unlikely to apply to real physical phenomena. Either hypothesis—God exists, or God doesn't exist—remains undecided when we assume a multiverse. A multiverse only makes the needed creator omnipotent on a much vaster scale. The multiverse and infinity take us to the realm of mathematics and its relationship with physics and cosmology.

# 9

## Mathematics, Probability, and God

**W**hat *is* the relationship between mathematics and science, and how does it impact the issue of God? And also: How does math relate to the reality we perceive in everyday life? These are among the most important questions one can ask about mathematics.

Many mathematicians are *Platonic* in their approach (after Plato's philosophy of ideal forms): they believe that numbers and mathematics as a whole have a life that is *independent of the real world*. Numbers, equations, and geometrical objects and ideas such as a perfect circle or a perfect square dwell in a milieu that exists independently of our universe. Some entity, an outside power—God—had to create all these concepts that are real regardless of human beings. Sciences such as physics, cosmology, chemistry, and biology *use* equations and other mathematical constructions from this Platonic universe. And these equations and other mathematical rules have to come from *somewhere*. If evolution created life, who created the mathematical rule of evolution? And even if the laws of quantum mechanics could make a universe, who established those mathematical quantum rules? The laws themselves can't just "evolve"—someone or something had to have *made* them.

**MATHEMATICS IS NOT** physics. Despite the historical relationship that goes back to Galileo and even earlier, there are fundamental differences between these two disciplines. The key difference is *reality*. In pure mathematics, we can create *any* reality simply by starting from some basic, and often seemingly arbitrary, assumptions. In geometry, for example, we define a point and a line.

From basic definitions, or axioms, mathematicians can build an entire theory by *proving* results—called theorems, or lemmas or corollaries (as results that precede a theorem or follow it are called, respectively)—using logical rules and techniques. But once a theorem is proved, there is nothing more to its veracity: it is known to be true regardless of any physical or other scientific reality or even metaphysical considerations. Once it is proved, a mathematical theorem just "is." And all future results will just *add* to it—they will never *replace* it.

Physics, and other branches of science, are different in this fundamental way from pure mathematics. In physics, a result believed to be true can always be replaced by a later one that may be found through theory and experimentation to better reflect reality. The replacement of the Ptolemaic model of the solar system by the Copernican model is the classic example in astronomy of one theory replacing an earlier, inadequate one. Similarly, Einstein's general theory of relativity replaced Newtonian mechanics as the theory of gravitation—although, when speeds are not close to that of light, and when masses are not as large as those of the sun, Newtonian mechanics still generally works very well. An example mentioned earlier is the problem of the regular shifting of the perihelion of the planet Mercury in its orbit around the sun, which can only be explained by general relativity. And it is worth mentioning that a GPS would not work properly without continuous electronic adjustments due to the effects of general relativity. So it is still true that general relativity has replaced Newtonian mechanics as the "true" theory of gravity—perhaps until an even more general theory combining quantum mechanics with general relativity is confirmed.

Furthermore, in physics, we require *experimental evidence* for all our theories in order to validate them. Generally, we build a theory, which typically will consist of a mathematical equation, or a set of such equations, with other parameters such as initial conditions. Then we use the model (the equations and additional parameters) to make *predictions* about new phenomena. If and when these predictions reflect accurately the results of experiments, we say that the theory is proved—until a time when a new theory may provide better predictions, in which case the new theory will replace the old one.

But mathematics has a very special—and rather mysterious—

relationship with physics, and to a degree also with other sciences. Mathematics seems to work exceptionally well in explaining the physical universe in ways in which a non-mathematical approach simply could not.

In his book *The Road to Reality,* Roger Penrose addresses the relationships among the three realms: mathematics, the physical universe, and the human mind. Penrose quotes Plato's famous Allegory of the Cave—the story of prisoners in a cave who cannot see its entrance or the real world outside, so must make deductions about it based on the shadows they see inside their cave. In a way, physical science today works in a similar way: what we observe and infer about nature is based only on the "shadows" that modern science can provide us about some hidden reality. We've discussed this idea in the context of quantum theory.

There have been quantum experiments, mostly about entanglement and various modern forms of the double-slit experiment, in which what the experimenter knows or *can know* does affect the outcome. For example, if detectors can be placed along the possible paths of a particle in a two-slit experiment, the particle will "choose" one of the two paths and there will therefore be no interference pattern on the screen. If no detectors are placed, so that the experimenter *cannot possibly know* which path will be taken by the particle, the particle will take *both* paths and interfere with itself. It might appear from this bizarre phenomenon that the mind somehow interacts with nature.

We don't know whether this is true, and many physicists do not believe in the existence of a connection between mind and physics—it is likely a spurious link that just appears to exist because of the way experiments are designed. But this is still an unsolved conundrum. Still, our mind *perceives* nature in a certain sense and to a certain degree and these perceptions are important to consider in any analysis.

These considerations have meant to Penrose that there are three "worlds" that mutually interact with each other: the Platonic mathematical world, the natural world, and the world that is in our minds, as represented in the figure below.

## ROGER PENROSE'S THREE WORLDS

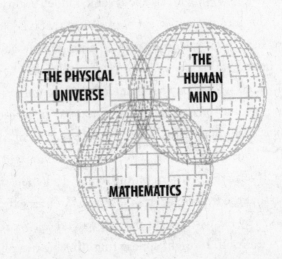

*Roger Penrose's three worlds, which mutually overlap:*
*mathematics, the physical world, and the human mind.*

Some of the purely mathematical world of ideas as defined by Plato reflects information about the physical world, although there are elements of the pure math world that have nothing to do with the real world; some of the physical world affects and is represented in our minds, and parts of it are inaccessible to human consciousness and understanding; and part of what goes on in our minds is accessible to representation in the world of pure mathematics, while other mental processes are perhaps non-mathematical.

Parts of mathematical truth are *inaccessible* to mental reasoning: the diagram "permits the existence of true mathematical assertions whose truth is in principle inaccessible to reason and insight"; there is a "possibility of physical action beyond the scope of mathematical control"; and there is allowance for "the belief that there might be mentality that is not rooted in physical structures."

Penrose is influenced in his thinking about mathematics by the work of the Austrian logician Kurt Gödel, who showed that in any system of mathematics there will always be propositions whose truth is *inaccessible* to us—we are unable to determine whether they are true or false (I will discuss Gödel's work later in the book). Penrose's conclusions are very profound and speak to the topic of this book and tell us that *science has limitations.* The limitations arise from the fact that perhaps not everything in nature is given to mathematical analysis, not all the content of mathematics is accessible to our mind, and the human mind itself may not be wholly reliant on and derived from purely physical or material notions.

The Nobel Prize–winning quantum physicist Eugene Wigner of Princeton wrote a landmark paper in 1960 that addressed the mysterious nature of the relationship between mathematics and physics and related sciences. Wigner had contributed much to physics using very advanced mathematics; in particular, he was one of the pioneers (with Hermann Weyl) of using abstract mathematical *groups* to model physical phenomena. Group theory is the mathematical branch dealing with symmetry, and as Wigner and Weyl have shown, symmetries can reveal deep secrets about physical reality.

Such symmetries, for example, allowed Steven Weinberg, another American Nobel laureate, to predict the existence of a particle and the actual masses of two particles: the so-called Z and W bosons, which act inside nuclei of matter to produce radioactive decay. The fact that the pure mathematics of group theory can produce such accurate physical predictions is remarkable. The connection between mathematical groups and physics was established by the German-Jewish mathematician Emmy Noether. Noether, who came to America just before the Second World War to escape Nazism, proved two major theorems that established the relationship between the symmetries of group theory and the all-important conservation laws in physics (such as the conservation of energy, which says that energy can neither be created nor destroyed).

In his 1960 paper, in which he marvels at the amazing relationship between mathematics and science, Wigner tells an amusing story about two childhood friends who meet after not having seen each other for decades. One of them had become a statistician and was proud to show his work to the other. Explaining the Gaussian (also called "normal") probability law, he showed him how an inference about some population of people can be made based on a random sample. The friend tried to understand, and pointed to a symbol that goes with the Gaussian integral. "What's this?" he asked. "Oh, it's pi," answered the statistician, "the ratio of the circumference to the diameter of a circle." The friend was incredulous: "Pi? But what does a population of people and a sample of people have to do with a circle?"

Wigner brings up this story as an example of the sheer mystery of why mathematics works so well in situations where some of its aspects (such as the constant governing the measurements of a circle) seem to have absolutely nothing to do with the phenomenon under study. And yet, it does! To be sure, the Gaussian curve is completely mysterious in other ways as well: it is one function that we don't know how to integrate (we can only do it numerically, using a computer, and not in "closed form" as a known function)—so although it is the most important function in all of probability theory, we cannot use the calculus of Newton and Leibniz to evaluate it.

Much of science deals with probability theory, and in order

to do science right, a scientist needs to have a firm grasp of it. In *The God Delusion*, Richard Dawkins makes an unsubstantiated statement: "The God Hypothesis is . . . very close to being ruled out by the laws of probability." Then, a few pages later, after he quotes T. H. Huxley:

> Huxley, in his concentration upon the absolute impossibility of proving or disproving God, seems to have been ignoring the shading of *probability*. The fact that we can neither prove nor disprove the existence of something does not put existence and non-existence on an even footing. I don't think Huxley would disagree, and I suspect that when he appeared to do so he was bending over backwards to concede a point, in the interests of securing another one. We have all done this at one time or another.

The misunderstanding of the modern theory of probability evidenced in this quotation also appears in Dawkins's discussion of a work by Stephen Unwin, *The Probability of God:*

> Unwin is a risk management consultant. . . . The plan is to start with complete uncertainty, which he chooses to quantify by assigning the existence and non-existence of God a 50 per cent starting likelihood each.

Dawkins is taking both Huxley and Unwin to task for giving God a 50 percent chance of existing (in Huxley's case, to "put existence and non-existence on an even footing," which is the same thing). Dawkins claims that such an assignment of prior probability is not in agreement with the theory of probability. Here is what is false about Dawkins's argument.

The great British statistician Harold Jeffreys, whose work laid the foundation for some of the modern theory of probability and statistical inference, published a book in 1939, titled *Theory of Probability*, in which he introduced the concept of a *noninformative prior distribution*. A noninformative prior is the only honest probability distribution one can use when there is no preexisting information in a statistical study. Furthermore, even when information exists, we really shouldn't use it in an a priori probability distribution if we want the data of a study to tell their own story in an unbiased way.

Jeffreys's formulation of honest statistical inference uses a term $1/n$, where $n$ is the size of the set of all possibilities. When we have two such possibilities: "God exists," and "God doesn't exist," we have $n = 2$; and therefore, by Jeffreys's rule, the correct a priori probabilities one should use in such an analysis are $1/2$ and $1/2$, or 50 percent each.

A noninformative prior distribution is one that is *flat*, with no peaks, because a peak would imply that the outcome corresponding to it has a higher probability than the rest. This is shown below.

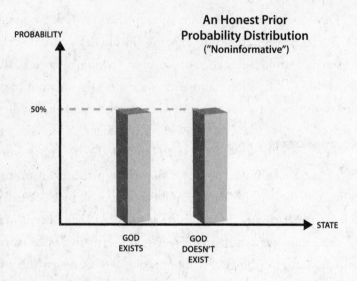

*The statistically honest prior probability distribution for the existence or nonexistence of God, using the standard Bayesian "noninformative" requirement for prior knowledge.*

Noninformative prior probability distributions are then used in Jeffreys's procedure for inference to result in unbiased conclusions based on the likelihood function that arises from the use of real data.

According to the eminent statisticians George Box and George Tiao:

> A noninformative prior does not necessarily represent the investigator's prior state of mind about the parameters in question. It ought, however, to represent

an "unprejudiced" state of mind . . . noninformative priors are frequently employed as a point of *reference* against which to judge the kind of unprejudiced inference that can be drawn from the data.

But no one would call Richard Dawkins unprejudiced. The only way to conduct a test for the existence of God is to start with the noninformative prior probability distribution that assigns each state, "God exists" and "God doesn't exist," an equal probability of 50 percent—exactly what Dawkins criticizes Huxley and Unwin for doing.

Dawkins claims to be a great admirer of the early-twentieth-century British evolutionary biologist and geneticist Ronald Fisher, the father of modern statistical inference. Fisher developed the modern theory of statistics while he was growing plots of tomatoes, and evaluating which of several fertilizer "treatments" worked best.

Fisher's work enables us today to determine which hypothesis among several is best supported by data, using the theory of probability. This process results in a key objective concept called the *p-value*. The p-value of a statistical test is *the probability* of obtaining the data we have, assuming the null hypothesis is true. A small p-value thus gives the conclusion of the test (the alternative hypothesis) high credibility. For example, if we conduct a test of whether smoking causes cancer, using a large group of randomly chosen smokers and nonsmokers, and conclude that smoking causes cancer with a p-value of 0.0001, it means that the chance that our result—our conclusion that smoking causes cancer—is wrong, based on our data, is only 1 in 10,000. This provides strong probabilistic evidence for our finding. A p-value of 0.1, on the other hand, is very weak because the chance that we are wrong in making the conclusion we have made is 1 in 10, and 10 percent is considered a high probability of having made a mistake.

Dawkins has not produced a p-value for his hypothesis that God, or some form of external force of creation, does not exist. So his conclusions ought not to be called science. He brushes off the fact that great scientists, Huxley in the nineteenth century and Stephen Jay Gould in the twentieth, admit that science and God may well coexist. Here is what he writes:

> Science can chip away at agnosticism, in a way that Huxley bent over backwards to deny for the special case of God. I am arguing that, notwithstanding the polite abstinence of Huxley, Gould and many others, the God question is not in principle and forever outside the remit of science. As with the nature of the stars, *contra* Comte, and as with the likelihood of life in orbit around them, science can make at least probabilistic inroads into the territory of agnosticism.

But *how*? Where is the probabilistic argument against God? Where are the likelihoods and where are the priors against the God hypothesis? In what way does science make its probabilistic inroads? There is a difference between finding out anything about the stars, or even discovering radio signals from distant civilizations (albeit something that has never yet happened) and disproving the existence of God. How does one discover probabilistic truths about God?

On the other hand, here is a real example of how p-values and good probabilistic arguments are presently used in particle physics:

The recent discovery of the Higgs boson, announced by the CERN laboratory, used the exacting level of proof required for all particle discoveries: a probability of 99.99997 percent, or a p-value of less than 0.0000003. Such a strict standard of proof requires an immense amount of data. And until such data were available, the careful scientists of CERN did not dare announce that they had "found the Higgs." Unlike these physicists, Dawkins has not tried to conduct a hypothesis test for the existence of God using *any* kind of probability.

**DAWKINS'S APPARENT UNFAMILIARITY** with the laws of probability extends to his misunderstanding of statistical issues—surprising when we consider how important statistics is for any branch of science, certainly for his area of expertise, biology. Here is what Dawkins says in describing a statistical study of Fellows of the Royal Society about their beliefs:

> All 1,074 Fellows of the Royal Society who possess
> an email address (the great majority) were polled,

and about 23 percent responded (a good figure for this kind of study).

The above is an excellent example of bias in statistical inference. The first thing any aspiring statistician learns is that we cannot trust any natural censorship: in this case, the use of e-mail. In an honest and rigorous statistical study, every member of the Royal Society would have to be approached in person, because e-mail *immediately* removes some people from the population of interest, creating a bias. (We all know that people respond to unsolicited e-mail messages in different ways.)

And while Dawkins says that a 23 percent response rate is a good result, one has to ask, good for *what*? This is a classical prescription for bias and misinformation. If only 23 percent respond to a survey, you can be sure that there is some biasing mechanism that is inherent here: either people of belief or nonbelievers are consistently avoiding a response—or else the response rate would not be so low. This is a case study of how *never* to conduct a statistical survey. In real surveys, one would return to the non-responders and try to obtain answers from at least some of them in order to assess the level of inherent bias, then try to correct it. As it stands, this survey is one of the shoddiest I have ever seen. If 78 percent of the responders said they did not believe in a personal God, it could still happen that within the overwhelming 77 percent of the sample that *didn't* respond, one might find large support for belief. This survey is worthless, and no self-respecting statistician would ever report its results.

Interestingly, religious and "pre-scientific" people living in the British Isles as long ago as the twelfth century were exceptionally successful—without the benefit of modern science—in coming up with a statistical methodology for testing the quality of gold and silver coins made at the Royal Mint in London. This story shows that with a lot of goodwill, hard work, and an effort to learn something about nature and the world, even deeply religious people can do the "right thing," which—given the time, the 1100s—should impress us today, when we know so much more thanks to the insights into statistics and probability provided by Fisher and others.

In Westminster Abbey there are several large wooden boxes for holding coins—each box called a *pyx*—from various centuries. (The word *pyx* comes from the Greek *pyxis*, which means box.) These boxes are historical souvenirs of the annual Trial of the Pyx, in which the Master of the Mint is "tried" by the Worshipful Company of Goldsmiths on behalf of the British Crown to see if he has responsibly handled his job and has not committed either of two errors: wasting the king or queen's gold by making coins that are larger than they should be or stealing some of the royal gold by making the coins smaller than required.

A coin is collected "at hazard"—meaning at random, something we are careful to do today in all statistical analyses—from all gold coins made during each day (the collection is called the *journee*, which comes from "day") and placed in the pyx. Once a year, as the box is close to being full, all the coins are counted and weighed. If the average weight of the coins is found to be larger or smaller by some preset amount than the standard weight of a coin, then the Master of the Mint is found guilty.

The statistician Stephen Stigler of the University of Chicago has studied the Trial of the Pyx and has concluded that even though the procedure is ancient, it still follows rules that are surprisingly close to those we would use today in a modern statistical hypothesis-testing procedure. The Trial of the Pyx shows us that some knowledge—a statistical understanding of nature that is mostly intuitive rather than mathematical—can also lead to excellent results.

Problems with probabilistic analysis arise immediately when one assumes an infinity of possibilities. The mathematical concept of infinity is very complicated, and we will come back to it later in the book. But I now want to address a simple property of infinity that is crucial for its implications about the multiverse and the anthropic principle:

> Given an infinite number of trials, any outcome that
> has a non-zero probability of happening will eventu-
> ally happen—and in fact will happen infinitely often.

Suppose that the probability of being run over by a car when crossing a street is extremely small; you can choose *any* number you like to represent it (as long as it is not zero, which would mean it *cannot* happen). Let's assume the probability is one in a billion. There is a rule in probability theory that says that, for independent events (crossing a street at different times can be assumed independent), you obtain the probability of being run over (technically, being run over at least once in as many trials as you make) as: one minus {the number one minus [the probability of being run over on a single crossing] raised to the power of the number of trials}.

You can experiment with various numbers of trials and a different probability of being run over on a single street crossing. What is important for us here is that any number less than one, when raised to an infinite power, will give the answer zero, and that then subtracting it from one will give one, which is a probability of 100 percent. If you try something infinitely often, no matter how unlikely the occurrence of the event might be, it *has* to happen.

In the famous popular example of the monkey typing *Hamlet*, it can be proven mathematically using the method above that a monkey sitting in front of a typewriter (or a computer keyboard) and *randomly* hitting the keys forever will, after an unimaginably long sequence of trials, type *Hamlet* (and all of Shakespeare's works, and all the books in any library). This will never happen in real life because the probability of getting all the letters of any work, such as *Hamlet*, correctly in a sequence if you're typing the letters randomly is vanishingly small. The heart of the argument here is the immense power of *infinity*.

If you allow infinity to enter any argument, anything can happen—even monkeys typing *Hamlet*. The play has about thirty thousand words, and if we assume an average of five letters per word, that is about 150,000 characters that the monkey needs to get right and in sequence. So the probability of getting it right the first time (leaving out spaces and punctuation, which would make it even more difficult) is one divided by 26 raised to the power 150,000, which is a number very, very close to zero—but not *identically zero*. Making the number of trials equal to infinity then "forces" the answer to be 100 percent. It is simply a mathematical fact that has no meaning outside the realm of pure mathematics and does not describe the real world in any way. So playing the "monkey typing *Hamlet*" game is not a good approach to real-life situations, and real universes—of which we know only one.

This is, in fact, the problem with the infinite multiverse favored by Brian Greene, Lawrence Krauss, Richard Dawkins, and others—it rests on a misunderstanding of the mathematical idea of infinity. For the minute you allow infinity to enter any argument, you can "prove" almost anything. If you assume an infinity of other universes, then simply by the immense power of the concept of infinity you can find a universe that will have all the required parameters for life to exist: the right masses and charges for all the elementary particles, the right balance of all the forces of nature, such as gravity and electromagnetism, and the right elements for life. If you have an infinite set of possibilities to choose from, of course one of them would be perfect for what you want. You need a multiverse in which all values on a continuum of values are present, so that you will find one in which, for example, the ratio of the mass of the proton to the mass of the electron is exactly 1,836.153 . . . (the actual, precise number that represents this ratio), just as it needs to be for matter to exist. And the same must be true for all other parameters of physics and biology. We see once again just how infinitely profligate the infinite multiverse model has to be.

And this isn't science, since it's not based on any reality, any experimentation, or even any *viable theory*. It is simply a "forcing argument" that allows you to prove anything you like. It's just like proving that a monkey can type *Hamlet* despite the unreality of the whole idea: the *only* reason it works is that infinity is such an overwhelmingly powerful concept. If you "go to infinity" (whatever that may mean, since infinity is inaccessible to us), you can pretend to prove anything. So the multiverse and the infinitely many copies of you and me that Brian Greene seems so eager to assume must exist out there (where, exactly?) mean absolutely nothing and really have no place in any scientific argument about nature, life, and our universe and where it came from.

But one can leave out pure mathematics and still see the main problem with "scientific atheism." Science as we know it today has been moving more and more into the realm of *information theory*. We now tend to view many things as pure information. To see how this works, consider an abstraction of what life involves: if you have a *rule* that tells you how to make a human being, this rule would consist of sheer information—the human genome, consisting of 3 billion bits of information, expressed as the letters of the genetic code, A, T, G, and C, arranged in pairs. So the question of God versus science boils down to the question: Who created the *information set* of life? Evolution certainly played a role, but then evolution itself is a code—an information set. And in order to "evolve," one needs to start with an initial construction, an initial set plus a rule of how it must evolve through time. Thus the question of a beginning of the process of life through a powerful set of commands, something like a primeval piece of computer code, is unavoidable. How is it possible to prove that no outside force created the code?

The same argument holds for the making of the inanimate world and the universe as a whole. Physics and cosmology, too, consist of a set of rules, a blueprint for creating a universe. There are rules of physics and initial parameters, and differential equations that tell a process how to start and how to proceed. And the code must include the masses and the charges and the strengths of all the forces of nature. It is a viable hypothesis that something—some outside power—had to have created the primeval information set that launched our universe on its way and eventually led to the emergence of life and intelligence and people asking questions about where we came from. But, as we see next, mathematics cannot always help us to fully explain the workings of the universe at a level that would allow us to make predictions about events and phenomena.

# 10

## Catastrophes, Chaos, and the Limits of Human Knowledge

The earthquake and tsunami that hit Japan in March 2011, damaging the nuclear reactors at the Fukushima Daiichi plant and thereby releasing large amounts of radiation, are an example of a *catastrophe*. Catastrophes are unpredictable events with undesirable consequences that buffet us every so often, causing havoc and devastation. They are unexpected, discontinuous in the mathematical sense, and often nonlinear. When events move along an expected path, progressing steadily from one point to the next, there are no catastrophes. By definition, a catastrophe is a *discontinuity* of a system as it progresses through time.

In the 1960s, the French mathematician René Thom derived a theory about abrupt, unforeseen change. Catastrophe theory, as it is called, is part of geometry and deals with dynamics in which a process has discontinuities: large shifts or fissures or jumps. The theory tries to model catastrophes—unexpected jumps in a system, such as Japan's 2011 earthquake and tsunami. A dog may stay quiet and well behaved for a long time, then suddenly, for no apparent reason, it will attack and maul a passerby. Or snow will stay piled up on the slope of an alpine cliff until, for no apparent reason, it suddenly turns into an avalanche. Earthquakes, tornados, hurricanes, and wildfires are all unpredictable catastrophic events. Small changes in plate tectonics take place over time, perhaps causing minor tremors from time to time, and then suddenly, without warning, a major earthquake occurs. Or the weather may be calm and clear one hour, and in the next, a tornado is destroying sturdy, century-old buildings.

While catastrophe theory attempts to explain the dynamics of such systems, the fact remains that we are often unable to predict a catastrophe. *Continuous* change—the change that is so well modeled by the calculus of Newton and Leibniz and is useful in explaining the manageable Newtonian universe—is easy to understand and to predict. If a variable is continuous, such as the distance you cover with your car as a function of the time you drive, huge, unexpected changes generally do not occur. But if a process is *discontinuous,* then no prediction is at all possible. Catastrophes demonstrate that for all our technological advances, there still are aspects of life and the physical universe that remain outside our ability to fully understand and control.

**MATHEMATICS SOMETIMES TELLS** us that we *cannot* know certain things. In chaos theory, we see how chaotic behavior, which cannot be predicted and is also *not random*, arises from very simple physical systems. Take for example a double pendulum made of two pieces of metal tied together by a string with the pivot point located in the middle, thus allowing the two lengths of metal to swing in a somewhat independent way: they will immediately develop unpredictable patterns of swing—completely chaotic motion.

Chaos is an extreme form of what we call nonlinearity. Linear variables grow slowly, as in a straight line. Nonlinear variables grow much faster and hence are less contained, less predictable, and less well behaved. For instance, if stock prices move proportionally to the third power of some economic variable, it means that doubling the economic variable's value will result in eight times the change in the value of the stock (because two to the third power is eight). This is an example of nonlinearity.

Turbulence is a highly nonlinear phenomenon. A hurricane is an example of how quickly turbulence can get out of hand. A storm may start linearly, in an ordinary, smooth pattern, become a tropical depression, and move along a path in the ocean. For reasons we don't fully understand, it may gather strength, and then the turbulent, highly nonlinear behavior of a hurricane starts and grows fast, feeding on itself. As the storm progresses, its wind speed increases, as does its energy and its power to destroy things in its path. Water behaves similarly to air when it comes to turbulence. Ocean currents, whirlpools, and other hydrological phenomena are very often highly nonlinear. The power of waves also can grow very fast in an unexpected and unpredictable way. A small change in one variable can create a wave or a current that is much greater in magnitude. Feedback mechanisms then make the phenomenon more and more uncontrollable, unpredictable, and violent.

Highly nonlinear mathematical systems are notoriously volatile. In order to understand how the chaotic double pendulum will swing, for example, we would have to know its initial conditions to a level of accuracy that is well outside our reach. Why? Because by definition, a chaotic system is one that *depends very strongly on initial conditions*. If you start at any *exact* point (and for us it is immaterial what the point represents—it could, for example, be the height to which you raise the double pendulum), say 7.512, then there will be a certain trajectory to the process that follows: a certain pattern for the pendulum. But if next time you start the same system at 7.5120000000000000000000001, a point that is extremely close to the previous starting point of the system, but not quite there, the trajectory of the system (in this example, the exact swinging mode of the double pendulum) will be completely different. We can prove mathematically that every time a chaotic system starts at a different starting point—even if that point is extremely close to the last one—the pattern of behavior will be completely new. Once again, we see that there are real limits to our knowledge of the world.

And unexpected, unpredictable, chaotic processes affect

everything in life—including the story of how we came to dominate the Earth. The impact on our planet of a large asteroid or meteorite 65 million years ago brought on the demise of the dinosaurs and the subsequent ascendancy of the primates, culminating in the evolution of humans. This impact was a discontinuous, abrupt event that brought on a decisive change. While the equations describing the motion of our planet and other solar system objects—if they were known—*might* have made this event predictable, the effect on our planet was not. And in fact, today we know that orbits of small objects in our solar system are sometimes chaotic in nature. The universe is full of such surprises, and they defy scientific predictability.

**SOON AFTER MY** book *Fermat's Last Theorem* was published, I received a call from Benoit Mandelbrot, the father of fractal geometry. Friends at IBM (where Mandelbrot then worked) had told me that mine was the only book that Mandelbrot liked so much that he carried it around with him wherever he traveled. We arranged a meeting at a small restaurant near my university. After a pleasant conversation about mathematics and life, I asked him: "How did you come up with the idea of a fractal?" His answer has intrigued me ever since: "From the stock market," he said.

Apparently Mandelbrot had looked at stock movements on varying scales: annual, monthly, weekly, daily, minute by minute, and instantaneous, and he noticed that they always looked the same—the jagged pattern of movement seemed identical, regardless of the scale. This gave him the idea of a self-replicating structure: a fractal. Such structures are intimately related to chaos theory. If you "live" on a fractal, your life is unpredictable—movements from point to point on the surface of a fractal are unpredictable and chaotic.

My point here is that even with the greatest mathematical knowledge we may possess, formulating equations as precisely as we can, there will always be variables and quantities that we will *never* be able to predict with any precision at all because they are chaotic in their nature.

The best example of this is of course the "butterfly effect," an idea conceived by the MIT mathematician and meteorologist Edward Lorenz. This is the hypothetical effect of a butterfly flapping its wings in China, unpredictably causing a hurricane off the East Coast of the United States. This is chaos theory at work: some extremely tiny disturbance in the air pressure in one place, constituting a minuscule change in the initial parameters of a system (here, the Earth and its weather), can cause a huge, unexpected change at a distant location.

Chaos theory demonstrates to us the limits of our knowledge: no matter how precisely we may think we understand a physical or natural system, there are things about it that we cannot know and cannot predict. To me, chaos theory represents a big and unavoidable hole in science that leaves us no choice but to admit the limits of our knowledge, our limits in explaining the workings of the universe. There is a huge variety of situations and circumstances in which systems become intrinsically chaotic or nearly so. As we have seen, it is often impossible to predict storms, earthquakes, tsunamis, and other natural phenomena. Sometimes these can be seen from the point of view of catastrophe theory, other times as intrinsically nonlinear events in which chaos theory explains the unpredictability of the event.

Stock market crashes, too, are highly nonlinear events that are related to chaos theory. The crash of 1929 is a case in point. The sharp jumps that Benoit Mandelbrot observed in the movements of stocks are occurrences that are unpredictable. A small change somewhere in the world economy can bring about a huge change in market values in ways we cannot even imagine—and even a collapse of the financial system.

After a day on Wall Street, the pundits all come out with explanations of what happened during the day: "The market fell because . . ." Or "The market rose because . . ." But these analyses are often meaningless, because the change that took place was unpredictable—and in this way also unexplainable.

Chaos theory, catastrophe theory, and nonlinearity teach us that some things are simply outside our understanding. Unfortunately, our brains are very linear in their thinking: we can follow a smooth path from point A to point B, but a highly unexpected, nonlinear trajectory is something we are not hardwired to understand or predict. And even our mathematics—usually reliable when describing a trajectory that is linear or relatively smooth—fails us here.

Our lack of ability to predict outcomes is not due to "bad mathematics"—it is a result of the fact that these highly complex systems are also hypersensitive to precise initial conditions: no one can predict how the butterfly's flapping wings will change the dynamics of a system. This quality of highly nonlinear systems represents a huge hole in our knowledge of nature and demonstrates once again that there are, and always will be, things we cannot predict in a satisfactory way. And what does this say about our ability to understand nature?

**IT IS IMPORTANT** to understand that chaos is *not* randomness. A chaotic system is perfectly nonrandom; it is *deterministic* in the sense that one outcome leads directly to another without any randomness, but we cannot know which one.

The fact that a system can be fundamentally nonlinear and unpredictable and yet not random is very important. It tells us that nature has processes and outcomes that are even outside of *probabilistic* analysis. Such things are intrinsically *unknowable* to us and, in a sense, lie in the realm of the gods—well outside human understanding and control.

What is so surprising is that chaotic, catastrophic, and highly nonlinear phenomena do not necessarily require many variables and inputs to get started. The fact that our double pendulum made of two bits of metal and a piece of string *immediately* exhibits chaotic behavior tells us that even very simple-looking processes can be unpredictable. And if we fail in our efforts to understand even the simplest of natural processes, how can we ever pretend to have knowledge that is so complete and so powerful that we can claim to have disproved the existence of God?

# 11

# Between God and the Anthropic Principle

This is the book's most important chapter, in which we consider the anthropic principle, which argues that the universe is the way it is because if it were any different, we humans wouldn't be here. The anthropic principle has been one of the most important tools in the hands of atheists in their battle against the notion of a God-created world.

The universe we see around us is characterized by extremely finely tuned constants—numbers such as the mass of the electron and the strength of gravity, on which the existence of our world depends. This has led some to believe that if we are here, then the world must be as it is. We have to live in the only universe hospitable to us so, within an infinite multiverse, we find ourselves in that universe in which we *can* exist. The anthropic principle, plus the existence of an infinite collection of (mostly inhospitable) universes, is seen by some as a good substitute for a God who purposely *made* the constants of nature what they are so that we could live.

As science broke new ground in the twentieth century, truths as strange as quantum theory emerged. Roger Penrose has spent a lifetime trying to understand the workings of the universe. And he has come to a stunning conclusion: if the entropy (a measure of disorder commonly used in physics) of space had been off from what it currently is by even a tiny fraction, the universe would not exist. Thus the universe has to have been "fine tuned" to a degree that we can hardly comprehend. Penrose writes, in *The Road to Reality:*

> Can the anthropic principle be invoked to explain the very special nature of the Big Bang? Can this principle be incorporated as part of the inflationary picture, so that an initially chaotic (maximum entropy) state can nevertheless lead to a universe like the one we live in, in which the Second Law of Thermodynamics holds sway?

The second law of thermodynamics says that the entropy of a system will increase through time. Penrose's model of a universe that gives rise to human life has certain requirements, such as the maintaining of the second law as well as conditions of equilibrium of temperatures and other variables that are consistent with it. He writes:

The argument would run roughly: "For sentient life to exist, we need a large universe with timescales long enough for evolution to take place, in conducive conditions, etc.; this requires some inflation, originating from our tiny smooth initial region, and once it starts, the inflation goes on to provide us with the wonderfully enormous observable universe that we know." Although it may seem that this picture is of such a marvelously romantic nature that it is completely immune from scientific attack, I do not believe that this is so. . . . The required precision in phase-space-volume terms is one part in $10^{10^{123}}$ at least. The exponent $10^{123}$ comes from the entropy of a black hole of mass equal to that in the observable universe.

Only a mathematical genius like Roger Penrose could come up with an argument for the existence of a life-giving universe based on the thermodynamic requirements of a black hole. Then Penrose refines his argument by asking: "But do we really need the conditions for life in the entire universe?" And his answer is that there is a minimum part of the universe that is forced to have benevolent conditions that could support life and intelligence. He can thus give up slightly on the requirements, leading him to conclude:

Thus, the precision needed, on the part of our "Creator" . . . to construct this smaller region is now only about: one part in $10^{10^{117}}$ Our Creator now only requires a rather *smaller* "tiny smooth region" of the "initial manifold" than before. The Creator is much more likely to come across a smooth region. . . . There was indeed something very special about how the universe started off. . . . We might take the position that the initial choice was an "act of God" . . . or we might seek some scientific/mathematical theory to explain the extraordinarily special nature of the Big Bang. My own strong inclination is certainly to try to see how far we can get with the second possibility.

Penrose draws a picture of "The Creator," a man with a long white beard pointing to an infinitesimally small point, within the entire "space of parameters" possible for the entropy of the universe, in order to have created the universe that we actually have. Not at all a religious man, Penrose nevertheless understands that something like a miracle might have created our world to the precise amount of entropy required for its existence. In seeking alternative reasons for this amazing cosmic "coincidence" of infinitesimally small probability, Penrose also admits that an ultimate theory of quantum gravity might someday lead us to another answer.

**PENROSE'S MOST FAMOUS** former student and scientific collaborator is Stephen Hawking. As we've seen, Hawking himself has hedged somewhat on the issue of the creation of the universe, sometimes choosing a more atheistic point of view than at other times. Throughout his life, it appears that Hawking, too, has been deeply concerned with anthropic arguments.

In September 1981, Hawking attended a conference at the Vatican. Addressing Hawking among a group of top scientists, Pope John Paul II said that it was probably futile for human beings to inquire into the actual moment of the creation of the universe. According to the pope, such knowledge comes "from the revelation of God." The pope was correct in pointing out that physics and cosmology are unable to bring us to the actual moment of creation, let alone take us beyond it to see what caused the Big Bang. So whether it was God or not, we are unable to explain the Big Bang. Sometime afterward, Hawking discussed this issue with author John Boslough, offering a telling glimpse of his view of the universe and how it might have come about:

> The odds against a universe like ours emerging out of something like the Big Bang are enormous. I think there are clearly religious implications whenever you start to discuss the origins of the universe.

Hawking has wondered about the parameters of the universe throughout his life and has pointed out that if the electric charge of the electron had been slightly different from what it is, stars would either not burn at all or would not have exploded in supernovas to spew into space many of the elements we need for life. If the force of gravity had been even slightly weaker, matter would not have come together to make stars and planets.

We have no theory that can predict why the forces and charges and masses are the way they are. These parameters seem arbitrary from a theoretical point of view. But without their values being precisely what they are, we wouldn't be here. Hawking also said the following: "If one considers the possible constants and laws that could have emerged, the odds against a universe that has produced life like ours are immense."

Hawking was also led to the anthropic principle in his attempts to explain how a universe supporting life, which a priori had such an incredibly small probability of emerging, ever came about. According to science biographer Kitty Ferguson,

> Hawking explains the anthropic principle as follows: picture a lot of different, separate universes, or different regions of the same universe. The conditions in most of these universes, or in these regions of the same universe, do not allow the development of intelligent life. However, in a very few of them, the conditions are just right for stars and galaxies and solar systems to form and for intelligent beings to develop and study the universe and ask the question, why is the universe as we observe it? According to the anthropic principle, the only answer to their question may be that, if it were otherwise, we wouldn't be around to ask the question.

Many physicists dislike the anthropic principle because it has no explanatory power, except for the trivial one that things are the way they are because they couldn't be any other way. And the anthropic principle is not a good replacement for God. You could say that there is one universe and that God made it this way—the parameters and forces all fitting perfectly well—so as to create intelligent life. Positing infinitely many universes and an anthropic principle to "choose" among them the one in which we must live is an unparsimonious way of building a model of life, and not a highly scientific one. Hawking and many other physicists hope that a "theory of everything" will some day *explain* the values of all the parameters of the universe so that the anthropic principle could be retired.

**ALTHOUGH THE ANTHROPIC** principle does not have the usual scientific validity or the power to truly explain things, the New Atheists embrace this theory because it is a substitute for God. Richard Dawkins devotes almost thirty pages of *The God Delusion* to this principle, curiously even linking it to natural selection: "Natural selection works because it is a cumulative one-way street to improvement. It needs some luck to get started, and the 'billions of planets' anthropic principle grants it that luck." Noting that the anthropic principle is "hated by most physicists," Dawkins says, "I can't understand why. I think it's beautiful—perhaps because my consciousness has been raised by Darwin."

Roger Penrose, in fact, sidesteps the anthropic principle.

For him, the origin of the universe is either an "act of God" or something that we may find when we have the "final theory" of physics. Like the multiverse, to which it is often linked, the anthropic principle is a kind of forcing argument that lacks a profound theoretical justification.

There are some variants of the anthropic idea. The *weak* anthropic principle guides variables such as why we are on Earth and not on Venus: Venus is too hot, so we must be here and not there. We must be within the "habitable zone" around our sun, a zone that satisfies the Goldilocks Quest—not too hot and not too cold; it is the region of space where water can exist in liquid form so it can support life as we know it.

The *strong* anthropic principle applies to everything: the masses and charges of all elementary particles, the cosmological constant, the entropy of our part of the universe, the strengths of all the forces of nature, and everything else. It says that all the parameters of nature are the way they are simply because if any of them had different values we simply wouldn't be here.

The anthropic principle has an interesting history. In the early 1960s, Princeton physicist Robert Dicke invoked what are essentially anthropic arguments to explain the age of the universe. He stated that the age must be compatible with the evolution of life, and, for that matter, with sentient, conscious beings who now wonder about the age of the universe. In a universe too young for life to have evolved, there were no such beings. But the term *anthropic principle* seems to have been coined in 1973 by the Australian physicist Brandon Carter, in a lecture he gave at a congress in Kraków celebrating Copernicus's five hundredth birthday.

Over the decades, Dicke's argument has been extended to other numerical measurements of the universe we observe around us, and thus to questions such as: Why is the mass of the proton 1,836.153 times that of the electron? Why are the electric charges of the up and down quarks exactly 2/3 and –1/3, respectively, on a scale in which the electron's charge is –1? Why is Newton's gravitational constant, $G$, equal to $6.67384 \times 10^{-11}$?

And there is also the question that has deeply puzzled so many physicists since 1916: Why is the *fine structure constant*, which measures the strength of electromagnetic interactions, so tantalizingly close to 1/137—the inverse of a prime number? (We now know it to far greater accuracy: about 1/137.035999.)

Richard Feynman once wrote: "It's one of the *greatest* damn mysteries of physics: a *magic number* that comes to us with no understanding by man. You might say the 'hand of God' wrote that number, and 'we don't know how he pushed his pencil.'" The astronomer Arthur Eddington (who proved Einstein's hypothesis that space-time curves around massive objects) built entire numerological theories around this number—all of them false. (He assumed that the constant was 1/136). There is even a joke that the Austrian physicist and quantum pioneer Wolfgang Pauli, who throughout his life was obsessed with the number 137, asked God about it when he died (in fact in a hospital room numbered 137) and went up to heaven; God handed him a thick packet and said: "Read my preprint, I explain it all there."

Stories and jokes aside, the values of all the physical constants described above have persistently defied all analysis or rational explanation. One physicist who made a strong attempt to understand them is Steven Weinberg, who seems to have always been ahead of his time. In 1998, just a few months before the announcement of the stunning astronomical discovery that the universe is accelerating its expansion—leading to the conclusion that "dark energy" permeates space, pushing the universe ever outward—Weinberg and colleagues at the University of Texas published a paper about that then-hypothetical dark energy. They argued that if it exists, its numerical measurement *must* fall within a very narrow range of values, which they specified in their paper; for otherwise, the energy would be too high for galaxies to coalesce through the gravitational force, or it would be too low and the gravitational force affecting all matter would win out, leading to a gravitational collapse before galaxies and life would have had time to evolve.

Weinberg and his colleagues thus derived what the value of the cosmological constant would have to be (within bounds) based purely on the anthropic principle. The anthropic principle helped predict the value of an unknown parameter, but the methodology used was not satisfying since it did not reveal any underlying reasons for the value of the cosmological constant other than "if we are here to observe it, it has to be within this given range."

Of course this argument would also apply to Newton's constant, the masses and charges of the quarks and the electron, the fine structure constant, the parameters governing the strong and weak nuclear forces, and so on. The forces of nature are *extremely* fine-tuned to accommodate a universe such as the one we see around us. Anthropically speaking, if we are here, the parameters have to be what they are.

The force of gravity, even though it is the one we feel the most, is in fact the weakest of the four forces of nature. Gravity is *forty orders of magnitude weaker* than the electromagnetic force. You can perform an experiment to prove it: Place a small paper clip on a table. The force of gravity, exerted on the paper clip by the entire planet underneath the table, is keeping it in place. Now take a small bar magnet and lower it down toward the paper clip. When you get close enough to it, the paper clip will jump up and stick to the magnet. This shows you that a very small magnet can *overcome,* using the electromagnetic force it generates, the gravitational pull on the paper clip that is exerted on it by the entire Earth.

Why is the gravitational force forty orders of magnitude weaker than electromagnetism? Why are the strengths of the four forces of nature exactly what they are? Without the highly fine-tuned values of the forces, we simply would not be here: gravitation would crush us before we had a chance to exist if it were any stronger, and if the electromagnetic force had a different strength, chemistry as we know it would not work because the electrical forces in atoms could not maintain the electrons in their orbits around the nuclei. If the strong nuclear force had a different value, quarks would be crushed or fly out of protons and neutrons, and atomic nuclei and therefore matter would not endure. And if the weak nuclear force had a different value, possibly almost everything would be radioactive, or stars would not shine and produce heat, so there would be no life.

When I interviewed Weinberg about his work and about the anthropic principle, he told me, "The universe could well be like a giant Schrödinger's cat. There are parts of the universe where the cat is alive, where the cosmological constant is just the right level and there are scientists there observing it and asking questions. And there are parts of the universe where the cat is dead—where the cosmological constant is too small or too large and therefore there is no life and no scientists asking questions about the universe." This is certainly one interesting view of the universe.

But the anthropic principle is used by some cosmologists simply because we do not know why parameters such as the masses and charges of the electron and the quarks, the entropy of the universe, and the strength of the cosmological constant are so immensely fine-tuned as to assure the existence of our universe.

If you wanted to test which hypothesis is true, a universe *created* to specific requirements, or a universe that just happens to satisfy the requirements because we observe them, you would find that there is no scientific way to determine the answer.

**AS WE HAVE** seen, physics is unable to escape the conundrum of the incredibly fine-tuned nature of many of its parameters. The best and simplest example of this mystery is that of the interplay of the proton, neutron, electron, and quarks. Every student of physics knows that matter is made of protons and neutrons inside the nucleus, orbited by electrons to complete the picture of the atom. Now, the dance of electrons around nuclei is achieved because the electric charge of the electron is equal in magnitude but opposite in sign to that of the proton: without this equality of charges, there would be no life-giving universe.

But while the electron is an elementary particle (it has no internal structure), the proton and neutron are not. Each proton is made out of three quarks—two "up" quarks and a "down" quark. So the electric charges of the quarks *must add up precisely* so that the charge of the proton will equal +1 (the electron's charge is defined as –1), or else the balance won't hold.

We know that this indeed happens: the charge of the "up" quark is exactly 2/3, and the charge of the "down" quark is exactly –1/3. When we add up the charges of the two "up" quarks and the single "down" quark that make up the proton we get 2/3 + 2/3 – 1/3 = 1. How could this happen so precisely? To further compound the mystery, the neutron (present in the nuclei of all elements heavier than hydrogen) must have an electric charge of zero, and it is composed of two "down" quarks and an "up" quark. And yet the mathematical magic works again. If you add the charges of the quarks that make up the neutron, you get 2/3 – 1/3 – 1/3 = 0.

Why would the charges of the quarks work out so perfectly? In the beginning, a fraction of a second after the Big Bang, the universe is believed to have consisted of a *quark-gluon plasma*, commonly referred to as a "quark soup." Then these quarks swimming in the dense and extremely hot soup created by the Big Bang suddenly bunched in *threes* to make protons and neutrons. This alone seems mysterious: generally in nature things pair up—not form threesomes. Why and how did all this happen, and how did the charges and masses and strengths of interactions for bunching up together to create stable composite particles all work out as needed to make a universe? Science has no good answers for these mysteries.

In fact, the standard model of particle physics was devised, using powerful mathematical ideas, to try to answer some of these very riddles—but it has absolutely failed to address the questions about the masses of the elementary particles and about the strengths of interactions of the forces, such as the infamous "1/137" governing all electromagnetic interactions. These numbers do not come out as results of the formulation of the equations of the model and have to be "put in by hand." Just how the "free parameters" in our models of the universe obtained the precise values they require so that our universe would exist remains a deep unsolved mystery—among the greatest riddles of science.

One way out of the problem is to say, "If the parameter values weren't what they are, we wouldn't be here to ask the question"—the anthropic principle. But one cannot use such a statement to scientifically falsify the competing hypothesis: "The parameters were created the way they are in order to make a universe." So is it God, or is it the anthropic principle?

**PERHAPS THE BEST** example of how inadequate the anthropic principle is in providing good, valid explanations for phenomena has to do with an event mentioned earlier, the crash into Earth of a large solar-system object 65 million years ago, which resulted in the atmosphere being filled with dust for years, causing freezing that killed much of life, including the dinosaurs. It is widely believed that had this event not taken place, dinosaurs would have continued to rule the Earth and primates would never have had the opportunity to evolve and eventually take over the planet as humans.

If a strict advocate of the anthropic principle were to be asked why an asteroid or meteorite hit the Earth 65 million years ago, his or her answer would have to be: "Because otherwise we wouldn't be here to ask this question." This answer is exactly in line with those to similar questions, seen throughout this chapter. But in this example we can clearly see why the anthropic answer is not science. A solar-system object hit the Earth 65 million years ago because its orbit happened to intersect that of our planet at that point in time. This is the correct *scientific,* non-anthropic explanation. It is immensely important to note, however, that no such explanation exists for the masses and strengths of interaction constants of the universe. And since the anthropic principle, as we see, is so unsatisfactory, one must consider other explanations. These may include divine intention, or at least something that resides well outside our present powers of understanding.

# 12

## The Limits of Evolution

When Charles Darwin proposed the principle of evolution in 1859, he forever changed natural science. Thanks to him, we now have an excellent scientific mechanism that shows how life-forms evolve and why better-adapted organisms gain an upper hand in propagating their genes. Evolution works on the principle of natural selection: better-adapted individuals are preferentially selected to reproduce.

As environments change, species that are better suited to them flourish, less-adaptable ones diminish, and some eventually disappear. Evolution thus explains why so many life-forms once existed and now live in the world.

The roots of evolution are in taxonomy: classifying living things into families, groups, and other classes, and seeing that a branching process governs their development. Analyses provide a clearer picture of how life evolved from simple to more complicated organisms and how different families of living things branched off. Such divergences might have been caused by mutations in genes that eventually created living things that were different from their ancestors and from nearby branches of animals or plants.

The gateway to Darwin's theory of evolution was opened in the eighteenth century by the Swedish physician, botanist, and zoologist Carl Linnaeus. While he was still a student at the University of Uppsala in the 1720s, Linnaeus wrote a paper about flower stamens and pistils that earned him an invitation to work at the university's botanical gardens. He excelled in his work, and in 1732 the Royal Swedish Academy of Sciences financed his trip to Lapland to study the flora.

His study of flowers in Lapland gave Linnaeus an outrageous idea: to try to *classify all the living things on Earth*. In 1735, after earning a medical degree in Holland, he presented his masterpiece: the *Systema Naturae*. By the time this guidebook to life-forms was in its tenth edition, two decades later, it contained information on and classifications of more than 7,700 plant and 4,400 animal species.

The Linnaean system said nothing about evolution. It was a system of classifying living things, and as such it implied that the world of creatures had an inherent order and that this order could be analyzed using the scientific method.

Another person paved the way for Darwin: the French scientist Georges Cuvier, who studied living and extinct elephant species and came to the understanding that fossils were the remnants of past living things. He concluded that species lived for generations and then, for unknown reasons, ceased to exist. In addition, a contemporary French biologist, Jean-Baptiste Lamarck, who had studied mollusks, worked out that creatures transmuted: living things evolved over time to become more complex. However, he did not believe that species extinction was possible.

All of these pre-Darwinian ideas were still contrary to a strictly literal interpretation of Scripture, according to which God created the living world, and its inhabitants did not change or evolve or become extinct, but rather lived static lives.

Despite these earlier ideas, when Darwin's *On the Origin of Species by Means of Natural Selection, or the Preservation of Favoured Races in the Struggle for Life* was published in 1859, it shocked most people because it seemed to call into question God's role as "creator," as well as the Genesis story. Even Darwin's wife felt uncomfortable with his work and expressed her fear that because of it she might never be reunited with him in heaven. But since many of Darwin's ideas were in the air, scientists and intellectuals had already become amenable to embracing them.

When he read Darwin's book, the naturalist T. H. Huxley reportedly exclaimed: "How extremely stupid of me not to have thought of that!" And eventually it was Huxley whose work on behalf of Darwin helped bring the ideas of evolution, adaptation, and natural selection to wider public understanding and acceptance.

**BORN IN 1809** in Shrewsbury, England, Charles Robert Darwin was the fifth of the six children of a wealthy country doctor, Robert Darwin, and his wife, Susannah Wedgwood, of the famed Wedgwood china-making family. Darwin's grandfather, Erasmus Darwin, had already proposed a theory, similar to Lamarck's, which he called *zoonomia*. It contained the germ of the idea of evolution even though it was unscientific and rife with inaccuracies and speculation.

Charles Darwin studied medicine at Edinburgh but reportedly didn't like it very much. He instead spent his time studying both Lamarck's work and that of his own grandfather. He transferred to Cambridge University to study theology, but ended up spending much of his time collecting beetles and other insects, riding horses, and hunting. He did well, however, in his study of natural history, botany, and geology.

Then, after his graduation, Darwin's botany professor recommended him for an unpaid position as a "gentleman's companion" to the captain of the H.M.S. *Beagle*, Robert Fitzroy. The ship was about to embark on a two-year mission to chart the coasts of South America. The professor thought this voyage would offer the young Darwin a wonderful opportunity to learn firsthand about natural history. As it turned out, the ship sailed for five years and went far beyond South America, eventually circumnavigating the world. This voyage changed the way we view the natural world, as it led to Darwin's great discoveries.

Throughout *On the Origin of Species*, Darwin deliberately

does not challenge religious beliefs. To the end, he is careful to allow for a "Creator," even though he avoids using a creator to explain things. Organisms evolve over thousands and millions of years from simpler forms to more complex ones. In the famous conclusion of his book, Darwin speaks about "several powers," leaving a place for a creator, but in a more passive way. Here is the final paragraph:

> It is interesting to contemplate an entangled bank, clothed with many plants of many kinds, with birds singing on the bushes, with various insects flitting about, and with worms crawling through the damp earth, and to reflect that these elaborately constructed forms, so different from each other, and dependent on each other in so complex a manner, have all been produced by laws acting around us. These laws, taken in the largest sense, being Growth with Reproduction; Inheritance which is almost implied by reproduction; Variability from the indirect and direct action of the external conditions of life, and from use and disuse; a Ratio of Increase so high as to lead to a Struggle for Life, and as a consequence to Natural Selection, entailing Divergence of Character and the Extinction of less-improved forms. Thus, from the war of nature, from famine and death, the most exalted object which we are capable of conceiving, namely, the production of the higher animals, directly fol-

lows. There is grandeur in this view of life, with its several powers, having been originally breathed ["by the Creator" was added here in the second edition] into a few forms or into one; and that, whilst this planet has gone cycling on according to the fixed law of gravity, from so simple a beginning endless forms most beautiful and most wonderful have been, and are being, evolved.

This was Darwin's summation of his book in one beautiful, poetic description of the process of evolution. Then the polemics began. T. H. Huxley (a.k.a. "Darwin's Bulldog") took up the fight on behalf of evolution, and his work is crucially responsible for the ultimate acceptance of the theory. Evolution sits well within biological considerations and can indeed explain a lot about the living world.

Many religious people—especially those who understand that the Old Testament was written for people who lived thousands of years ago and who could not possibly have had the scientific sophistication we have today—accept it as an excellent, viable theory to explain how complex life-forms came about.

According to the late, great evolutionary scientist Stephen Jay Gould:

Darwin's chief quarrel with creationism resides not so much in its provable falseness, but in its bankrupt status as an intellectual argument—for a claim of creation teaches us nothing at all, but only states (in words that some people may consider exalted) that a particular creature or feature exists, a fact established well enough by a simple glance: "Nothing can be more hopeless than to attempt to explain the similarity of pattern in members of the same class, by utility or by the doctrine of final causes. . . ."

But while Gould strongly believed that evolution did not agree with any creationist claims whatsoever, he was open to religious interpretations that accepted evolutionary theory, despite being himself an atheist. In this context, he says: "Darwin, for example, and following Hutton, Lyell, and many other great thinkers, foreswore (as beyond the realm of science) all inquiry into the ultimate origins of things."

This approach, of course, leaves open the issue of how life first came about. In his book *Rocks of Ages,* Gould says that the question of God's existence is outside science, a view held by other scientists as well but abhorred by Dawkins. Responding to Gould's statement, Dawkins writes:

I simply do not believe that Gould could possibly have meant much of what he wrote in *Rocks of Ages.* As I say, we have all been guilty of bending over backwards to be nice to an unworthy but powerful opponent, and I can only think that this is what Gould was doing. It is conceivable that he really did not intend

his unequivocally strong statement that science has
nothing whatever to say about God's existence. . . .
It implies that science cannot even make *probability*
judgments on the question.

As we have seen before, when Dawkins doesn't like what
another scientist says, he sometimes dismisses the statement by
claiming that the scientist "didn't really mean it" or said it "for a
purpose," such as to appease his or her opponents. It is also un-
clear how Dawkins wants to construct honest probability statements about the existence of God. Where should the objective
probabilities come from? There is simply no scientific method
for doing this in an unbiased way.

But it is clear why Dawkins would criticize the evenhanded
Gould. Here is what Gould wrote at the beginning of his *Rocks
of Ages*:

I am not a believer. I am an agnostic in the wise sense
of T. H. Huxley, who coined the word in identifying
such open-minded skepticism as the only rational position because, truly, one cannot know. Nonetheless,
in my own departure from parental views (and free,
in my own upbringing, from the sources of their re-
bellion), I have great respect for religion. This subject
has always fascinated me, beyond almost all others
(with a few exceptions, like evolution, paleontology,
and baseball).

Even a great modern evolutionary biologist such as Gould accepted the value of religion in our lives, which is evident throughout his book. A conflict between evolution and religious belief is only mandatory in the eyes of the New Atheists, and therein lies the true source of Dawkins's enmity. For Dawkins, even according respect to the ideas of religious people is a sin. He devotes a whole section in *The God Delusion* to stating his reasons why those who believe in God deserve no intellectual consideration from the enlightened, such as himself.

But evolution and belief in God are not necessarily contradictory. Even early in the twentieth century, the prominent French philosopher, geologist, paleontologist, and Jesuit priest Pierre Teilhard de Chardin was able to argue convincingly that evolution does not replace God. The creator—or whatever you want to name the force that brought us here—could well work through evolution.

"God acts through the process of evolution," Teilhard said. "I see no contradiction between evolution and my faith in God." For him, the laws of evolution themselves had to have been "created," one way or another. In this view, just because evolutionary processes exist, they do not replace an original creator who has set evolution in motion and created the germ of life. Teilhard was as devout as can be and yet wholeheartedly believed in science and in what it teaches us about the world.

He was one of the paleontologists who in the 1920s were involved in the great discovery of the fossilized remains of Peking Man—one of the "missing links" between humans and apes—in the cave at Zhoukoudian, southwest of Beijing, which caused great excitement in the world of anthropology. Teilhard was in China at the time because he had been exiled there by order of the church, whose officials were uncomfortable with his writings and lectures promoting evolution. Ironically, they had banished him to the one place in the world where he could do them the most damage, by being involved in "practical evolution" through work on analyzing the Peking Man finds, which date from about six hundred thousand years ago.

Peking Man was a member of a hominid species called *Homo erectus,* which preceded modern humans and Neanderthals and had a cranial capacity between those of the apes and ours. Teilhard's work proved scientifically that Peking Man made and controlled fire for cooking and for heating caves.

**THE THEORY OF** evolution has flaws. It does not explain a host of behaviors and phenomena. Why would certain animal species exist at all? My favorite example is the peacock: Is its huge, cumbersome tail really necessary for attracting females? These peahens must be more demanding from their males than Hollywood celebrities. Wouldn't a smaller tail do? Doesn't this exuberant tail make the animal immensely vulnerable to predators? How does evolution allow it to exist, let alone require it for the survival of the fittest?

The problem of altruism is perhaps the most hotly debated issue in evolution today. A charitable person who makes an anonymous contribution to help the needy does not gain any evolutionary advantage whatsoever. So, from a purely evolutionary point of view, as society moves forward in time, eventually traits such as altruism should disappear—but they don't.

The story becomes even more complicated when we add in what we have learned from the study of modern genetics. Natural selection tells us that individuals strive to propagate their genes through future generations. Therefore, risking one's own life to save complete strangers from a burning house should be a tendency that would disappear from the world: by doing this, a person increases the probability of removing his or her genes from future generations.

So it seems that evolution loses out as a principle. One explanation that has been offered is that the courage such an individual shows acts as a peacock's tail in attracting females. So the seemingly altruistic fireman enjoys a prestige that helps him attract healthy, beautiful women with great genes, so once he survives the risk of dying, he ends up propagating his genes better than men who are less willing to risk their lives.

Another route to explaining altruism is to argue that it still saves a person's genes, in that altruistic people will tend to rescue their own kin: children or relatives. And in doing so, they will indirectly be promoting their genes in future generations according to a formula derived by the evolutionary biologist W. D. Hamilton: save two of your children at the expense of your own life; or four of your cousins; or eight of your second cousins, and so on—when you do so, statistically you are saving your own DNA. Already in the 1930s, the British geneticist

J. B. S. Haldane said he would "lay down my life for two brothers or eight cousins." However, would people die for a sibling more willingly than for a child? Such "genetic" computations seem questionable.

And in fact, how would one use evolution to explain the persistence in our world of people such as soldiers, medics, firefighters, police officers, and first responders who courageously save members of a different race than their own at the risk of losing their own lives? There are so many such examples in our lives and our history. European Christians saved the lives of Jewish children in the Holocaust at the very grave risk of being executed by the Nazis. White civil rights activists in the 1960s risked their own lives to help blacks in the U.S. South. Soldiers speak of a bond of brotherhood that develops in combat—often leading to selfless acts of courage on behalf of one another. To claim that these are acts that propagate one's own genes would seem preposterous.

The American geneticist E. O. Wilson has recently backed off from his lifelong belief that altruism could be explained by endowing evolutionary advantages. Wilson chose a different approach in his newest book: likening people to a colony of ants. Ants will "choose" to assume roles that may shorten their lives so that the entire colony would survive. But in the case of people, it's not at all clear that this happens. What is the human analogue to the "colony"—is it a family, a platoon, a community, a nation, or the human race?

How often do we hear about a person who jumped into the

icy water of a lake to save the life of a dog, or a fireman who returned to a burning house to rescue a cat? Neither will be beneficial to the "colony"—even assuming it is the entire human species, in Wilson's analogy. So truly altruistic behavior—which we know has existed in the world since time immemorial—is not well explained by evolution.

An example that, to use one of Dawkins's favorite expressions, "deals a knockout blow," to evolution as an acceptable explanation for altruism is the following. In May 2012, several groups of individuals attempted to climb Mount Everest, as happens every spring. One of the climbers, a young Israeli, was well on his way to the top when he came across a fallen Turkish climber who had lost his face mask, his oxygen supply, and much of his equipment. He was clearly going to die very soon because of the extreme cold and oxygen-poor air near the summit.

The Israeli, who was in excellent shape and about to achieve a lifelong dream he had been training for all his life, stopped immediately and spent several hours helping the disabled Turk get back down the mountain. He saved the Turkish climber's life, at the loss of three of his own fingers and four toes to frostbite, and also lost his chance to make it to the summit of Everest.

Turkey and Israel have been political enemies since the Turkish-led flotilla to Gaza in May 2010. There is no love lost between these nations. The Turkish flag, with its star and crescent, was visibly sewn on the climber's outfit, so the Israeli saw it. Other climbers had already given the Turk up for dead. The Israeli climber would have gained far more sexual-selection kudos had he made it to the top than by bringing down a dying Turk and losing fingers and toes. So why did he do it? And why do many other people perform such acts of immensely courageous altruism at great personal loss to themselves—never accounted for by saving any part of their own, or their kin's, DNA? The answer is simple: human decency and goodness. There is no scientific explanation for such behavior from the field of evolution.

**DANIEL DENNETT IS** a leader in the atheism movement in America. He and his wife organize and participate in cruises for atheists, as well as other events aimed at cementing a worldwide community of atheists. But these atheist groups do not get involved in charitable work, as many religious organizations do. While staying at a hospital, you may be visited by a nun, a rabbi, or an imam, bringing you food or newspapers or comforts of many kinds, as many people know from personal experience. And it is true that sometimes religious people do this in an effort to convert patients—but not always. There are innumerable examples of religious persons who engage in charitable work in hospitals, poverty-stricken communities, halfway houses, and the like out of sheer charity and the urge to do kind things for complete strangers. I have never heard of an atheist group volunteering to offer comfort to the ill or the distressed.

**THE TWO THEORIES** that have been used by New Atheists to argue against the existence of God have been quantum mechanics in the case of the physical-cosmological realm and evolution in the biological one. These two theories are very different from one another, and at this point it may be instructive to compare the two.

## Comparing Evolution with Quantum Mechanics

| Quantum Mechanics | Evolution |
| --- | --- |
| Extremely mathematical | Non-mathematical |
| Makes verifiable (probabilistic) predictions | Makes no reliable predictions |
| Governed by precise formulas | Little use of formulas |
| Verified to a stunningly high accuracy | Accuracy is unknown |
| Its natural underpinnings are unknown | Describes nature directly |
| Very precise | Somewhat general |
| Complicated principles | Simple-looking principles |

Evolution by natural selection explains the richness of life we see in the world today, as well as all past branches of evolution. No serious scientist doubts that evolution is real and that it describes the emergence of new life-forms on Earth. We know that life probably started with very simple, single-celled organisms many millions of years ago, and through the processes of Darwinian evolution evolved into more complex, multicellular life-forms.

The idea behind evolution seems simple, and as the table above shows, it does not depend on complicated mathematical formulations. The fittest survive—those individuals that are best suited to their environment. Because they survive best, and are perhaps the healthiest and the most attractive to the opposite sex, they are the ones that tend to reproduce most successfully and hence propagate their genes in the population. Natural selection means that nature, rather than an outside entity, "chooses" the best individuals, i.e., the best-adapted ones, the ones that live and survive the best, to promote their genes. They mate, pass on their "good" genes to the next generation, and then the same forces of adaptation and natural selection do their magic again, naturally choosing the best individuals, who reproduce better, produce better-adapted offspring, and so on.

Nature has evolved attractiveness criteria for creatures. It would appear that the peacock's magnificent but cumbersome tail would make it difficult for such a bird to survive. But evolutionary biologists argue that the tail is a mechanism for sexual selection: the healthiest, strongest, most fit males are the ones that have the most impressive tails. Because their tails are attractive, they get to mate more often and with healthier females, and hence the beautiful, elaborate tail is passed on from generation to generation.

As you can see, the theory may *explain* what we see in nature, but it does not make good *predictions*, as quantum theory does. We cannot predict what species will inhabit the Earth a thousand years from now. And we could never, ever have predicted that an animal with a huge ornamental tail beyond any reasonable proportions would ever have existed.

In contrast, quantum mechanics is a theory we likely are in the early stages of understanding, but it produces exceptionally accurate probabilistic predictions of future outcomes. Galileo taught us that "the book of nature is written in the language of mathematics," and indeed our best theories are mathematical. The fact that evolution suffers from a paucity of mathematical concepts underlying its structure is something we should be concerned about. And perhaps this is why the theory has many holes in it; it may well be an incomplete theory in its present state.

So while it is ignorant and unscientific to fail to recognize that evolution is a powerful principle that often explains what we see in the biological sphere, it is equally unjustified to assume that evolution is a perfect theory that explains everything. A theory that cannot produce excellent *predictions* of future outcomes and phenomena is not a complete theory.

Unlike quantum mechanics, evolution has many variations. For example, Gould favored a theory in which changes in a species take place quickly, followed by periods of relative stability.

Others believe that evolution is slow and continuous. Dawkins thinks that intelligence arises naturally, and therefore must eventually evolve in any circumstance, while Gould believed that intelligence is a rare fluke of evolution. There are simply too many things we don't know about evolution—so to claim that it explains everything by natural selection is at least premature.

**DAWKINS IS PASSIONATE** in his belief that God does not exist. I asked him about it: "If you feel so strongly that any religion can't possibly be right, why did you say, in advertisements you paid to be placed on buses in London, 'God *probably* doesn't exist, so just enjoy your life'?" Dawkins's answer surprised me: "I thought it would be funny," he said.

Science is dispassionate, rational, a logical search for facts and truths about nature and the universe around us. It is the pursuit of the laws of nature, with no agendas to push for any philosophy about who created these laws. But the New Atheists, who claim to speak for science, are more like religious evangelists bent on converting us to their narrow point of view that God does not exist.

We don't know how the first living organisms came about, before they evolved into higher creatures. And we don't know the process that led to the emergence of eukaryotic cells—a huge advance over the earlier, simpler single-celled organisms. (Eukaryotic cells are those that have complex structures inside such as mitochondria, are protected by a membrane, and contain their genetic material in a nucleus.) The emergence of such a sophisticated cell structure, present in all advanced life-forms, is not fully understood or explained through evolution.

This advance was the key to the development of the rich life-forms we see on Earth. And we have no idea how such sophisticated cells became a reality. There are deep mysteries that have not yet been solved by evolutionary theory. Dawkins, for one, is not concerned with the questions that evolution has not been able to answer: how life originally arrived on Earth, how eukaryotic cells came into existence, and how intelligence and consciousness developed. What he says is that there are gaps in our knowledge, and that God should not "live inside the gaps." He adds, somewhat disingenuously, that God is too great a concept to be "relegated to the gaps." But it depends on your point of view: Is evolution a half-empty or half-full glass? Possibly, the "gaps" are the key elements here, while evolution itself is simply the icing on the cake.

If continuous movement is the norm in nature, then nonlinear or chaotic behavior can be seen as an example of "gaps." But, once again, it is these chaotic phenomena that may be key to the mysteries of nature. The gaps may be the important elements that we can neither explain nor ignore.

But next I will discuss an example where evolution may not even address satisfactorily a phenomenon about our own emergence as the dominant species on Earth—let alone predict it, and let alone fill in any gaps.

# 13

## Art, Symbolic Thinking, and the Invisible Boundary

Advances in paleontology and physical anthropology since the nineteenth century have brought us close to an understanding of how the human species has evolved over many millions of years—from fish to reptiles, to early mammals, to primates, and finally to a hominid that was remarkable seemingly only for its upright gait. This is indeed an accomplishment; however, these advances have been seized upon by New Atheists to explain away the remarkable arc of human history, whereby we became the creators of breathtaking technological and cultural achievement totally unique in nature.

IN 1974, THE American paleoanthropologist Donald Johanson discovered in the Afar region of Ethiopia one of the most important "missing links" between humans and apes: a three-

and-a-half-foot-tall hominid he called Lucy (so named because the Beatles song "Lucy in the Sky with Diamonds" happened to be playing on the tape deck around the time of the discovery). Lucy is part of a genus of hominids called *Australopithecus,* which have been discovered at various sites in Africa and dated to between 3.9 and 1.7 million years ago. As time progressed, these hominids walked more and more upright (relying less on their arms, as evidenced by their posture and the shortening of the arms) and their cranial capacity increased as they approached human form.

The fossil record, collected, organized, and studied over more than a century by dedicated physical anthropologists and archaeologists, reveals the story of the emergence of human beings on planet Earth. Our common ancestors with chimpanzees seem to have lived in Africa around 7 million years ago. Successive hominids then evolved away from our ape relatives. *Australopithecus ramidus* is the first of them, dated to about 5 million years ago, followed by *Australopithecus anamensis,* found in the Turkana region of Kenya, with a larger cranial capacity, living some 4 million years ago. Then came *Australopithecus afarensis,* found in the Afar region of Ethiopia—Lucy being the prime example of this hominid—walking more upright and with a still larger cranium.

The australopithecines (as these species of hominid are called) diverged from our ancestors about 3 to 4 million years ago and there emerged a species called *Homo habilis,* whose fossils were discovered in Tanzania by Mary and Louis Leakey in the 1960s. With a cranial capacity of six hundred cubic centimeters, *Homo habilis* had less than half our brain size, but twice that of the more advanced australopithecines. This was a hominid that lived roughly 2 million years ago (estimates range from 1.7 to 2.3 million years), stood about four feet three inches tall, and could make stone tools (hence the *habilis*, meaning able or handy).

Then came a very hardy and wide-ranging species called *Homo erectus* (upright-walking man), a creature that lived in Africa and in many locations in Asia, including Java and China, and is known to have used fire (as Teilhard de Chardin helped prove) and to have made more advanced stone tools than before. This hominid lived from around 1.5 million years ago till about six hundred thousand years ago and had a cranial capacity of more than one thousand cubic centimeters. Peking Man, an example of this species, lived in caves southwest of Beijing about seven hundred thousand to six hundred thousand years ago.

*Homo erectus* is an ancestor of modern humans through a number of intermediaries, prime among them *Homo heidelbergensis,* whose remains have been found in Germany as well as in Africa. *Homo heidelbergensis,* which lived between five hundred thousand and three hundred thousand years ago, had about our cranial capacity (with a range of from eleven hundred to our average of fourteen hundred cubic centimeters) and made stone tools. This primate is believed to have been the ancestor of both modern humans and Neanderthals.

What do we learn from the fossil record? We clearly see here an evolution over time that is characterized by increasing size—from three feet to close to six feet in height over several million years; the improved ability to walk upright; better manufacture and use of stone tools; emergence of the exploitation of fire; and a marked increase in the size of the brain. What do these trends mean? Humans clearly were not "created" in one act by God. They certainly evolved from our common ancestors with the apes to more and more advanced creatures that become more like modern humans. And at some point in time—one that we have not identified—we "become" human. Becoming human entails a mode of symbolic thinking and making art, as evidenced through the cave paintings of early humans who lived in Europe in the Paleolithic era.

Earlier hominids, while they fashioned stone tools that allowed them to kill and butcher animals, did not create art—at least not much of it. Every once in a while, a stone artifact that looks like it might have been carved as a primitive statuette is reported to have been found, and there is some ambiguous evidence of the manufacture of paint, but we have no definitive proof that any of the hominids that preceded us on Earth possessed an artistic ability.

Only our species is able to make and interested in making images of what we see around us. An artistic sense seems to be one aspect that separates us from all other living creatures and those that have inhabited the Earth in the past (perhaps with the exception of Neanderthals—that question is still open). And this artistic capacity represents our unique ability to think *symbolically*—something that our nonhuman ancestors probably could not do. Our art is very old: at least as old as forty thousand years, the age of the Paleolithic paintings that have been discovered in the El Castillo cave in Spain.

**WHEN DID CONSCIOUSNESS** arise, of the kind that we humans have? Are hominids such as Lucy more like apes or like us? And when does self-awareness arise in the animal world in general? Where is the invisible boundary in evolution at which an animal-like creature becomes humanlike? We have hardly any answers to these questions.

This notion of *emergence* is one that has been addressed in philosophy, but never explained well by science: We don't know how a universe emerged. We don't know how from the chaos and fuzziness and unworldly behavior of the quantum, the structured universe of macro objects we see around us came about, with its causality, locality, and definiteness—none of which are characteristics of the quantum realm. We don't know how self-replicating life emerged from inanimate objects. And we don't know how and why and at exactly what point in evolution human consciousness became a reality. The inexplicability of such emergent phenomena is the reason why we cannot disprove the idea of some creative power behind everything we experience around us—at least not at our present state of knowledge.

Here is what Richard Dawkins says about the emergence of human consciousness:

Imagine that an intermediate species, say *Australopithecus afarensis,* had chanced to survive and was discovered in a remote part of Africa. Would these creatures "count as human" or not? To a consequentialist like me, the question doesn't deserve an answer, for nothing turns on it. It is enough that we would be fascinated and honored to meet a new "Lucy." . . . Even if a clear answer might be attempted for *Australopithecus,* the gradual continuity that is an inescapable feature of biological evolution tells us that there must be *some* intermediate who would lie sufficiently close to the "borderline" to blur the moral principle and destroy its absoluteness.

Dawkins does make an interesting point: to whom do we accord "humanness"? But he skirts the main issue: To what extent can evolutionary theory answer this question? Evolutionary science cannot indicate to us the location of the point on the continuous evolutionary scale, which Dawkins believes is there, at which human consciousness arises. Evolutionary theory is unable to tell us how life began, how eukaryotic cells evolved, how intelligence came about, or how consciousness arose in living things.

The question about consciousness is key to everything we are discussing. Modern cognitive science relies on the principles of evolution and posits that consciousness is something that can be produced artificially. Life-forms become more and more advanced through evolution, and eventually consciousness is the outcome. Thus, many cognitive science practitioners believe that machines can develop a consciousness as well, although this has never happened. Consciousness has never been produced in the lab, not even close.

Their thinking is that in the same way that a computer can be "taught" to play chess so well that it will defeat a human, the machine can actually be taught to think and feel like a human being. But despite formidable attempts by computer scientists and cognitive science experts, they have not accomplished this. We can make a computer do many things, but we cannot make it react like a person—with consciousness, including self-awareness and free will.

**AT THE 2011** Ciudad de las Ideas conference, I was pitted (together with Dinesh D'Souza and Rabbi David Wolpe) against the evolutionary psychologist Robert Kurzban, the cognitive psychologist Gary Marcus, and the skeptic Michael Shermer. We were taking part in a debate on whether life has a meaning. My opponents argued that meaning is something that can be created artificially and that in fact a machine can create meaning for itself. Thus, they argued, there is no design or purpose in our universe. "We are all robots," Robert Kurzban announced.

I argued that we humans have never been able to create consciousness, and that therefore it does not follow that consciousness is likely to just arise on its own in a machine. We can build more and more powerful computers and robots—but these computers and robots will not have free will or self-awareness. The other side maintained that eventually they will. But if we have not ever built a machine that developed consciousness, then how can we claim that life and consciousness and free will can be developed in a lab? And without anyone being able to exhibit how these qualities that make us human come into existence, "scientific atheism" cannot prevail.

**IN HIS BOOK** *The Singularity Is Near*, Ray Kurzweil fancifully imagines a future civilization in which computers that have developed a consciousness build even more powerful computers for their own needs. But this scenario is fictional. We have not created even a shadow of consciousness in any machine thus far. Consciousness, symbolic thinking, self-awareness, a sense of beauty, art, and music, and the ability to invent language and pursue science and mathematics—these are all qualities that transcend simple evolution: they may not be absolutely necessary for survival. These attributes of the human mind may well be described as divine: they belong to what is way above the ordinary or the compulsory for survival. The origins and purpose of consciousness and artistic and musical and literary and scientific creativity remain mysterious. Why would evolution alone bring about such developments that appear to have little to do with the survival of an individual or a species?

The problem with consciousness is that we don't really understand what it is. Daniel Dennett writes in his book *Consciousness Explained* that many experts from a variety of fields such as psychology, anthropology, neuroscience, and artificial intelligence have been working on trying to understand what consciousness is, and he notes wryly: "With so many idiots working on the problem, no wonder consciousness is still a mystery."

In his research, Dennett pursues what he sees as Descartes's dream of understanding consciousness and defining a strict set of logical rules that govern the human mind—something he admits has been a lifelong professional dream. He summarizes his theory about how humans got their consciousness as follows:

> There is no single, definitive "stream of consciousness," because there is no central Headquarters, no Cartesian Theater where "it all comes together" for the perusal of a Central Meaner. Instead of such a central stream (however wide), there are multiple channels in which specialist circuits try, in parallel pandemoniums, to do their various things, creating Multiple Drafts as they go. . . . The seriality of this machine (its "von Neumannesque" character) is not a "hard-wired" design feature, but rather the upshot of a succession of coalitions of these specialists. The basic specialists are part of our animal heritage. They were not developed to perform particularly human actions.

Dennett and his collaborators consider the human mind from two problematic viewpoints: looking at the brain as a kind of computer, and looking at the brain as the result of animal evolution. The human brain is far more than a computer: computers have no consciousness. And to think of the brain as simply something that has evolved out of animal ganglia and primitive brains is also a mistake: there is a giant leap from the brain of a monkey or a dog to the brain of a human being.

Neither approach explains Leonardo's *Mona Lisa*, Picasso's *Guernica*, Beethoven's Ninth Symphony, or the palaces on Venice's Grand Canal. Neither do they explain Einstein's general theory of relativity or Freud's invention of psychoanalysis. Both the mechanistic and animalistic views of the brain fall flat in their attempts to explain any of these great historic achievements of the human mind. We are not machines, and we are not simple animals, either.

An alternative explanation is that God gave us the mental abilities and that extra something we use in making decisions and in creating great works of art, sublime music, magnificent architecture, beautiful literature, and science and mathematics. Our incredible brains can do all these things because they contain some ingredients that science has not yet found or explained and whose origin remains one of the deepest mysteries in all of science.

In his book *The Selfish Gene*, Richard Dawkins says that machines can have consciousness, or at least can act as if they had consciousness:

Each one of us knows, from the evidence of our own introspection, that, at least in one modern survival machine [by which Dawkins means humans], this purposiveness has evolved the property we call "consciousness." I am not philosopher enough to discuss what this means, but fortunately it does not matter for our present purposes because it is easy to talk about machines *as if* motivated by a purpose, and to leave open the question whether they are actually conscious. These machines are basically very simple, and the principles of unconscious purposive behavior are among the commonplaces of engineering science.

Once again, with absolutely no scientific justification whatsoever, a New Atheist wants to reduce the amazing human mind with its hopes, desires, aspirations, abilities, creative genius, goodness, love, and other complex emotions and qualities to a simple machine. Dawkins further writes:

In the chess-playing computer there is no "mental picture" inside the memory banks recognizable as a chess board with knights and pawns sitting on it. The chess board and its current position would be represented by lists of electronically coded numbers.

When Dawkins's book was originally written, the computer had not yet beaten a grandmaster, but Garry Kasparov's loss to Deep Blue in 1997 proved that a computer could prevail over the best chess-playing human. Dawkins predicted this would happen based on the fact that computers had been winning against successively better players. But to assume that the human brain, with its consciousness and everything that makes it unique, can be likened to a machine—powerful as that machine may be in its ability to manipulate large amounts of data at an amazing speed and using an immense memory capacity—is wrong. And let's not forget that a human is programming the computer!

**WE ARE FACED** here with one of the greatest unsolved mysteries in the history of science: At what point in hominid development and evolution does human consciousness appear? What exactly is our consciousness? What makes us different from animals? What gives us the powers to create and to think symbolically and to develop language?

The magnificent European cave art we find in Paleolithic caverns in France, Spain, and Italy gives us one of the earliest glimpses of the evolution of consciousness and symbolic thinking. According to the paleoanthopologist Ian Tattersall, symbolic thinking is the single most important characteristic of human beings, the one that separates us from all our predecessors, ancestors, and other animals. Symbolic thinking allowed human beings to create amazing art many thousands of years ago. It brought us language, science, art and everything that makes us uniquely human. Neither computers nor animals can do any of these things. So the emergence of consciousness and symbolic thinking remain one of the most formidable hurdles in the path of atheism. We have no good explanation of how consciousness and symbolic thinking came about. These may well be described as divine gifts.

# 14

## Engaging the Infinite

The New Testament book Ephesians identifies God as "Creator of all, who is over all and through all and in all." The verse succinctly captures the fundamental understanding of many faith traditions: the world was created through divine oversight, and God is infinite and all encompassing ("through all and in all"). Eager to topple these pillars of belief, New Atheists have countered that the universe was self-generated from nothing and that the universe, not God, is infinite. However, "nothingness" and "infinity" are well investigated mathematical concepts, and a deeper look at the mystery of numbers reveals how very far the New Atheists have to go to disprove God.

**LAWRENCE KRAUSS COMPLAINS** in his book that the "theologians keep changing the definition" of emptiness every time he "proves" that the universe came out of nothing without a divine creator. But as we shall see, it's Krauss who does not want to address the real concept of pure emptiness—because it's a serious challenge to his position.

In set theory, the foundational structure of mathematics, there is one basic concept called the *empty set,* also known as the null set. The empty set is that which contains *no elements at all.* This is as close as the human mind can come to understanding nothingness. Pure nothingness is defined as the contents of the empty set. This unique set contains nothing at all: no space, no time, no directions, no elements, no forces, no substance, no ideas, no notions, no rules. Nothing!

If you want to get a feel for this kind of complete and unqualified emptiness, draw a circle as a "something"—it contains a piece of space: it is a something, not a nothing. Now, shrink this circle slowly until it becomes a single point. Once the point is the only thing left, erase it. Now you have emptiness: not a piece of paper—that is gone too, along with the space, directions such as up and down and left and right, and any elements whatsoever.

Such complete emptiness means nothingness—no space and no points and no directions. From such a complete emptiness, no universe can ever arise. It is here that Krauss and others go wrong. Pure emptiness is something so deeply devoid of any entity that nothing can come out of it. In the Vilenkin model, on which Krauss's universe-from-nothing theory is based, a preexisting quantum foam must exist. And yet the mathematical empty set is far emptier than the quantum foam: it stands for complete and unqualified emptiness. It is a "nothing," while the quantum foam is, of course, a "something," leaving open the big question of how the universe and its particles of matter were ever created.

**WE NOW MOVE** to discussions of the math behind the science, and address what is and what is not knowable. One of the most interesting results in the area of pure mathematics called set theory is Russell's paradox. Early in the twentieth century, the renowned British philosopher and mathematician Bertrand Russell was able to show that there is *no set containing everything*. But before we get to this difficult logical paradox, it is useful to look at a simplified one, often called the Barber of Seville paradox.

The barber of Seville is known for shaving all the men in the city who do not shave themselves. So, does the barber shave himself? If he does (as a man of Seville), then he doesn't (as the barber); and if he doesn't (as a man of Seville), then he does (as the barber). This is a paradox—it has no solution. There simply can't be a barber satisfying these exacting criteria.

Russell's paradox is similar, but deeper. It seeks to answer the following question: Is there a universal set—i.e., a set that contains everything? To logically address this problem, Russell divided sets into two kinds: sets that contain themselves as an element, and those that do not contain themselves. The set of all dogs is not itself a dog. So the set of all dogs does not contain itself. But the set of all things that are not dogs does indeed contain itself (since it is not a dog). Now Russell asked the question: Consider the set of all sets that do not contain themselves. Does this set contain itself?—if it does, then it doesn't; and if it doesn't, then it does. This proves that there is *no set that contains everything*.

This logical paradox may well have deep implications about the universe, as we see if we ask ourselves the questions: What is the universe embedded in, and what is that embedding part of? Where do these embeddings stop? The paradox tells us that there is nothing that contains *everything* inside it. So what *is* the universe, what contains it, if there is nothing that is allowed to contain *everything*? The implications about what is knowable to us about the universe are disturbing and hint at our inherent inability to ever know everything about creation. This notion will become clearer soon.

Set theory was derived and extended by a tormented German genius, the mathematician Georg Cantor, who died in a sanatorium in 1918. Cantor spent his life trying to understand infinity. He felt that this endeavor got him closer to God, who he believed held the key to the deep truths about infinity that he was after. Cantor derived some immensely important facts in pure mathematics: he discovered that there are various *levels* of infinity, and he even learned how to carry out arithmetical operations on infinite quantities.

It's said Cantor had such a deeply perceptive mind that he could, in a sense, "see" infinity. He was the first mathematician in history to truly elucidate the deep properties of the infinite, all on his own, to a surprisingly great extent. He was able to demonstrate definitively that not all infinite quantities are equally large. For example, the number of integers, though certainly infinite, is *smaller* than the number of all the numbers that are found on the real number line (a set that includes not only all the positive and negative integers and all positive and negative ratios of integers—which Cantor showed had the same "size" as that of the integers—but also all *irrational* numbers such as pi and *e*), called the *real numbers*. It is the far more numerous irrational numbers that give the real number line its "substance" or density.

The real numbers are "infinitely dense"—between any two of them, no matter how close they may be, there is yet another number, and another, and another. . . . Thus, for any number, there is no "next" one, because if you choose a "next" one, there are still infinitely many numbers between the given number and the one you've designated as "next."

Now you can see why the notion of the infinite multiverse— an invention so favored by New Atheists—is absurd. The existence of a multiverse is exploited to enable one to "find" the one particular universe within the infinite collection that by chance alone satisfies the requirements for matter as we know it and for life (because we know that the parameters of our universe seem to be very finely tuned for our existence). For this, one needs a continuum of parameter values from which to choose ours, because the parameters of the universe are "exact" numbers, like pi or *e*, so that they naturally live on the continuum, and the continuum has a very high order of infinity (even if one argues, as one might, that any number on the real line can be approximated to any given level of accuracy using rational numbers, whose cardinality—the size of their infinite set—is that of the integers, as Cantor proved, it still leaves us a daunting infinity of universes).

But where *are* all the "infinitely densely packed" other universes, which are needed for this choosing mechanism to work? These universes must be as "close" to each other as the points on the real line (or at least as close as the rational points on the line are to one another). The mathematical situation is physically hard to imagine. If so many other universes existed out there, surely at least some of them would by now have bumped up against us?

Cantor was often harassed by less gifted mathematicians who found his work too bizarre to believe. Their constant attacks on him contributed to his mental illness. He suffered from recurring bouts of depression, which sometimes landed him in a mental institution where he would spend months until he felt better and was released. This cycle of productive and frenzied work under adverse conditions, alternating with periods of hospitalization and rest, characterized much of his life.

The conflict took religious overtones. Cantor's main adversary was the Berlin mathematician Leopold Kronecker, who used to taunt Cantor by saying: "God made the integers, and all the rest is the work of man!" He did not believe that numbers such as pi or *e* (which live on the continuum) exist. The conflict was a philosophical one because infinity is such an "unreal" concept. Cantor just happened to have a deep, intuitive understanding of infinity that transcended pure logic. For most of us the concept is too immense to truly comprehend.

Today we know that Cantor's work was perfectly correct and extremely innovative, and it has opened up important new ways of thinking about the infinite. But for all his success, there was one conundrum Cantor could not solve. Called the "continuum hypothesis," it states that there is no set whose cardinality is strictly between that of the integers and that of the real numbers. Resolving it seeks to answer the question of how many levels of infinity there are between that of the integers and that of the numbers on the real line.

Cantor's quest was really to understand the meaning of space and to determine its components. As such, his thinking went back to the same problem addressed by Leibniz a century and a half earlier when he invented his monads as the basic building blocks of both space and spiritual and metaphysical reality. Cantor, too, had a spiritual approach to the problem, in addition to the purely mathematical one. Cantor was a deeply religious man, a Lutheran with Jewish roots in Saint Petersburg and Denmark, and he believed that God "told" him that the continuum hypothesis was true, meaning that after the infinity of the integers (and fractions) comes the infinity of the numbers on the real line.

We don't really understand what space and its infinitely many points really are. This fact bears strongly on physics and cosmology because it implies that the actual space in which we live—far from being a "nothingness"—possesses a mysterious, deep structure. Since, as we will see, we can never know whether the continuum hypothesis is true using our mathematics, we can never hope to fully understand the nature of space. Physicists are not fully aware of the degree to which unsolvable problems in pure mathematics impact what we can ever hope to learn about space and time, the universe, and its origin.

To come close to a better understanding of the universe, we would need to know whether mathematical space is an accurate description of real physical space-time, or whether there are some other possibilities that may describe it. It is conceivable, for instance, that quantum effects make space and time "granular" (meaning made up of very small discrete entities, like sand) rather than continuous. Either way, contemplating space and its points brings us to the realm of the infinite, a concept we are unable to fully grasp.

Philosophically, one could say that the infinite belongs to God—humans cannot perceive or understand the infinite in any meaningful way, despite progress in mathematics by Cantor and those who followed him. To religious thinkers, as to Cantor, God *is* the infinite—something that exists but that we cannot reach or adequately describe or understand.

While he was under house arrest at Arcetri, Galileo spent time contemplating the infinite and arrived at a deep truth: that the order of infinity of all the positive integers was the same as the order of infinity of all the squared integers, as Galileo could show by matching 1 with 1, 2 with 4, 3 with 9, and so on. The fact that an infinite set can be put into a one-to-one correspondence with a *proper subset* of itself is, in fact, the property that *characterizes* infinite quantities. This should give you an idea about just how bizarre and incomprehensible infinity really is.

A simple, amusing example exhibiting this unexpected property of infinite sets is the thought experiment called the Infinite Hotel or Hilbert's Hotel (after the great German mathematician David Hilbert, who described it): Suppose that after a long and tiring flight you arrive in a strange city and discover to your chagrin that there are no hotel rooms available. Finally you notice that there is in this town a hotel called the Infinite Hotel, so you go there and make one last effort to find a room for the night. You ask the person at the reception if there is a room available—certainly an infinite hotel has infinitely many rooms, you tell him. But the attendant shakes his head. "Sorry," he says, "we do have infinitely many rooms, but unfortunately all of them are taken." You ponder this answer for a minute: infinitely many rooms, and yet every single one of them is taken. Then you have a sudden idea. "Look," you say, "I really need a room. Would you do me a favor?" "I'll try," says the receptionist. "OK," you say, "move the person in room number 1 to room number 2, then move the person from room number 2 to room number 3, the person in room number 3 to room number 4, and so on to infinity. In doing so, you will have freed up room number 1 for me."

This story demonstrates the incredible property of infinite

sets: here you match all the numbers from 2 to infinity with all the numbers from 1 to infinity, showing that there is the same number of numbers in both sets (that number is "infinity"—or rather, the lower order of infinity characteristic of the integers and the rational numbers).

Cantor named his orders of infinity, the infinite cardinal numbers, the *alephs*. The term may have come from his Jewish roots: in Kabbalah, God is denoted as the infinite, and the first letter of the word infinity in Hebrew is *aleph*. Cantor's investigations of infinity were intriguingly similar to what the Kabbalists had done in a non-mathematical way in trying to understand the properties of infinity in order to try to learn something about God.

In the case of the continuum hypothesis, mathematics fails: we are unable to truly understand the actual levels of infinity that exist and how they relate to one another. Possibly, if mathematical analysis fails, one might resort to metaphysical notions. And in fact, this is what Cantor did. He was hampered in his lifelong efforts to fully understand infinity, so he turned to a kind of deep spirituality, in which he saw God telling him that the continuum hypothesis was true. To Cantor, God was actually the culmination of all the alephs, a level of infinity that was so large that it was *inaccessible* (through any mathematical operation, including exponentiation) from any lower level of infinity. The concept of God as the highest possible level of infinity is beyond our mathematical abilities to comprehend, Cantor deduced.

It seems that every time Cantor spent much time attempting to prove the continuum hypothesis, he eventually succumbed to depression. A brilliant mathematician, he felt that he *should* be able to prove the hypothesis. In him we see the convergence of three entities: mathematics, spirituality, and the human mind. All three reflect on the universe and on what it means. Mathematics has proved a surprisingly powerful tool for analyzing and learning about the real world. Spirituality exists in the realm outside logic, mathematics, and science. And the human mind is what allows us to contemplate everything around us.

The tormented Cantor felt that he should be able to use pure mathematics and its logical laws and tools to answer the question about infinity and about the nature of space. His approach had earlier led him to an entire "paradise" of unexpectedly powerful results in mathematics. As David Hilbert described it: "No one will expel us from the paradise that Cantor has opened for us." Hilbert also presented, at an international congress of mathematicians held in 1900, Cantor's continuum hypothesis as the first of ten (later expanded to twenty-three) problems in mathematics that he hoped would be solved in the twentieth century. To this day, the continuum hypothesis remains unproved: we still don't know what space is really made of and what its structure, in terms of infinite quantities, truly is.

I bring up Cantor not only because we are addressing the idea of infinity, which bears on our discussion of the structure and origin of the universe, but also to provide an example where strict physical-logical analysis fails. Physicists are users of mathematics, not makers of mathematics. The truths used in physics are derived—often in greater generality and power than is necessary for applications—by pure mathematicians. And mathematicians work differently from physicists. They use logic, but they sometimes might also use intuition and feeling. Often a mathematician will "see" or even dream an important result before proving it rigorously.

We think of the universe as governed by strict logical laws, but in fact quantum theory, and ideas in pure mathematics, are not based only on logic. Cantor's work was governed by psychology almost as much as logic. It is here that we see the human mind transcending the rational and the straightforward. Our minds are based on essentials that go beyond the mechanistic and the evolutionary: they have something extra that allows them to do amazing things that computers, and dogs and monkeys, cannot. I believe that this mysterious extra element inside our brains—such as the ability Cantor possessed for dealing with the immense concept of infinity—is related to the divine.

**IN 1937, THE** brilliant (and, like Cantor, mentally troubled) Austrian logician Kurt Gödel was able to show mathematically that within our mathematics we are *unable* to either prove or disprove the continuum hypothesis. (Gödel's proof was completed by Paul Cohen of Stanford in 1963). This means that some truths about infinity *can never be known to us*. The infinite is so profoundly complicated that, hard as we may try, there are things about it that will always remain beyond our reach. This statement is not a guess or a hypothesis—it was mathematically proved and is accepted as true by all mathematicians.

But Gödel went deeper, and he produced one of the most profound mathematical results ever derived. Gödel's incompleteness theorems state that there are properties of numbers that will forever remain outside our reach: we will never be able to conclude definitively whether or not they are true. He also showed that our system of mathematics cannot prove its own consistency. Philosophically, Gödel's theorems place limitations on human knowledge—they demonstrate mathematically that some truths are outside our knowledge, and must remain so.

If applied to science, the implication of Gödel's theorems is clear: we will never be able to know *everything* about our universe because we are part of it. Thus Gödel's theorems could imply that we may never be able to decide the question of the existence of God.

Why is this so?

Throughout the history of science, mathematics has been our tool for understanding nature and its laws. To explain gravitation in a non-relativistic way, we use the mathematical laws of the calculus developed by Newton and Leibniz. Mathematics gives us a complete understanding of gravitation—up to the ability to land a space probe on Mars—when speeds are not too close to that of light. When speeds are great or gravitational forces are immense, the unique mathematics of Einstein's general relativity (such as the absolute differential calculus, the work of Gregorio Ricci-Curbastro and Tullio Levi-Civita, tensors, and Riemannian geometry) gives us the perfect answers. In the realm of the very small, the mathematics of quantum theory (called Hilbert space methods) gives us excellent answers—if not a real understanding of what is going on. But where do we go from here? What mathematics can explain the deepest laws of the cosmos? (Some might say string theory, but it hasn't provided complete answers so far.)

Gödel, like Cantor before him, was motivated and aided by abilities that went beyond simple logic. Sure, he was one of the greatest logicians of all time—perhaps the greatest—but his personality, his psychology, his sensitivity, and his intuition all played important roles in his mathematical work. In his biography of Gödel, *Logical Dilemmas: The Life and Work of Kurt Gödel*, John W. Dawson writes:

Many of Gödel's interpretations—of historical events, films, literature, political and economic affairs, and seemingly mundane happenings—struck his contemporaries as far-fetched or even bizarre. In his mathematical endeavors, however, his willingness to countenance possibilities that others were wont to dismiss or overlook served him well. Unlike Russell, for example, he took seriously Hilbert's idea of using mathematical methods to investigate metamathematical questions.

We see here a great mind, fundamentally different from others and able to use both strict logic outside of mathematics and some extralogical tools within mathematics. Gödel thought about the universe in a unique way, and it enabled him to prove that there is an end to what is knowable to human beings.

Spirituality, at least in an abstract sense, also played an important part in how Gödel thought about the world. Dawson, comparing Gödel's thinking with Einstein's (Einstein and Gödel shared a warm friendship while both worked at the Institute for Advanced Study at Princeton in the 1940s and 1950s) and with Leibniz's, wrote about Gödel:

> He shared Einstein's belief that we live in an ordered universe, created by a God who "doesn't play dice"; he embraced Leibniz's visions of a *characteristica universalis* and a *calculus ratiocinator*. . . . His consummate faith (so often vindicated) in the power of his own mathematical intuition led him to search for axioms from which the Continuum Hypothesis might be decided, to find a consistency proof for arithmetic based on constructively evident, though abstract, principles, and to expect that astronomical observations would eventually confirm that the universe in which we live is a rotating one of the sort he had envisioned.

Gödel was Platonic—he believed that numbers and other mathematical entities have an existence of their own, independent of the physical universe. On August 26, 1930, at the Café Reichsrat in Vienna, Gödel discussed his ideas about the incompleteness of mathematical systems with a number of logicians and intellectuals, including Rudolf Carnap and John von Neumann. In his notes about the meeting, Gödel wrote:

> It was the anti-Platonic prejudice which prevented people from getting my result. This fact is a clear proof that the prejudice is a mistake.

The Platonic point of view is consistent with the belief that some deep structures had to be created (or at least exist) that transcend the material universe. In the Platonic philosophy of mathematics, numbers and other elements of pure mathematics have an existence all their own. They do not "evolve" or come about "out of nothingness"—they exist, and are perhaps a product of some infinite wisdom or force or entity that permeates the universe and goes beyond it. But mathematics also holds the key to coming to terms with the physical universe and its laws.

On the other hand, mathematics shows us its own limits. By Gödel's incompleteness theorems we know definitively (Gödel *proved* his theorems) that some truths about a mathematical system—and this could include models of the physical universe—must remain outside our reach.

Because of the assumption that God may well "reside" outside the universe we inhabit, it stands to reason that the question of the existence of God may be one of those Gödel-like mathematical truths that will forever stay outside our sphere of knowledge. We don't know that this is the case, because we have no deep understanding of the universe's inception and what preceded it. But because we already have to admit a complete lack of knowledge about the structure that led to our universe's emergence 13.7 billion years ago, it is likely we will be prevented from ever gaining definitive knowledge about God—we may never know whether God exists or not.

When I interviewed Nobel laureate Steven Weinberg on the subject of human knowledge about the universe and its laws, he told me, "I don't know if the human brain can attain full knowledge of the universe—but I hope that it can. Maybe it will take a thousand years. . . . The Greeks predicted the existence of atoms, and it took two thousand years and we now know that atoms exist." The laws of nature may well be decipherable, as Weinberg hopes. But the existence of God may be a far more complicated issue, one that may very well lie outside the realm of science, and be mathematically impossible to address.

**OUR DISCUSSION OF** mathematics and infinity points strongly to the possibility that the greatest mystery of all—Is there a God?—may be one of those truths that are unattainable to us from a purely logical-mathematical framework. We see from work in cosmology that even the greatest theoretical physicists cannot bring us knowledge about what preceded the universe or answer the mathematical question, What set contains our universe? There are a multitude of other questions about our existence and about what the universe consists of and how it ever arose that may be mathematically impossible to determine. We can argue forever about God's existence, but the truth may very well be outside our ability to obtain.

# 15

## Conclusion: Why the "Scientific" Argument for Atheism Fails

The prolific eighteenth-century Swiss mathematician Leonhard Euler, one of the greatest mathematicians of all time, was also a deeply religious man. He was a member of the Royal Academy of Sciences in Saint Petersburg, Russia, and one day, the famous French atheist Denis Diderot came to visit the academy, apparently on a mission to convert its members to atheism.

Euler was told about the visitor, and believed that Diderot knew nothing about mathematics. So he surprised him in a public debate by demanding, "Sir, $a$ plus $b$ to the $n$th power divided by $n$ equals $x$; therefore, God exists! Respond!" Diderot, understanding nothing, could not open his mouth. Wild laughter erupted in the room; the humiliated Diderot withdrew and, the next day, packed his bags and returned to France.

This story may well be apocryphal, but what the New Atheists are doing today is tantamount to the same thing. Without a shred of evidence on their side, they declare: "Science proves there is no God! Respond!"—and a public that may not be steeped in all the nuances and technicalities of science is left flummoxed, and hence vulnerable to the New Atheists' overconfident pronouncements.

As we have seen through our discussion of mathematics, physics, cosmology, biology, genetics, brain and cognitive science, and evolution, science has severe limitations when it comes to determining the existence of God. Mathematically, it has been proved definitively that there are always facts within any mathematical structure that will remain forever outside our understanding, outside our knowledge, outside our reach.

In physics and cosmology, despite all our efforts to explain the values of the constants of nature by using all kinds of theories, we are unable to explain even conceptually simple properties of the physical constants needed for life to occur in the universe. This is a huge failing of science, because when models of the universe are built, the hope is always that they will lead to an understanding in the form of predictions of the values of the parameters of the theory. Our physical theories have thus failed us completely in this task. Some values of constants in quantum mechanics, quantum field theory, and relativity have been predicted, but most of the key physical properties of nature—the masses of the elementary particles that make up the universe and the strengths of interactions of the four forces of the physical universe—remain outside our understanding. Why is the fine structure constant—which governs all the electromagnetic interactions in the universe—equal to about 1/137? Nobody has even a clue. And as we've noted again and again in this book, the same is true for many other key constants of nature.

In the face of such limitations, physicists and cosmologists have been forced to retreat from their goal of reaching a full understanding of nature, and instead some have chosen an unscientific and unattractive explanation: the anthropic principle. This theoretical copout consists of pretty much raising your hands in the air and saying: "Well, if the constants of nature weren't what they are, we wouldn't be here." Of course doing so leaves most scientists with a deep sense of dissatisfaction. Einstein would likely have frowned on this principle since his life's goal was to unravel the laws of nature and come up with theories that *explained* why the constants were what they were based on the theories themselves, and to use the theories to *predict* what these constants should be. Instead, we are left with a disappointing void.

We see, therefore, that we lack a deep understanding of the workings of the universe. There are things that we know, and science has indeed brought us great truths. But we don't know what caused the Big Bang. We don't know how the molecules of life first arose on the surface of our planet. We don't know how the advanced cells of life came about, necessary ingredients for the evolution of complex organisms such as us. And we don't know the origins of intelligence, self-awareness, symbolic thinking, and consciousness. We lack basic knowledge about the most important and most enduring mysteries of creation.

And even if we could somehow obtain all knowledge about the universe, we could not possibly go beyond it—see *behind* the structures that science exposes, so that we could determine how the universe was "made." These inherent limitations in the very nature of science and knowledge make it unlikely that we will ever be able to solve the problem of God. At any rate, we have not solved it yet. And given all the power and complexity and depth of modern science, we have not been able to overrule scientifically the hypothesis of some form of external creation.

**WE DON'T KNOW** how or why our universe was created 13.7 billion years ago, and probably never will know. A tiny fraction of a second afterward, a primeval, mysterious, abstract symmetry suddenly broke through the action of a hidden field called the Higgs field, and mass was created—necessary for building a universe.

The *superforce* that emerged out of the Big Bang successively broke down into the four separate forces we perceive in the universe today: gravity, electromagnetism, the weak nuclear force, and the strong nuclear force. The magnitudes of these forces seem to have been perfectly designed for what would follow. As the universe expanded, it underwent a phase of immense growth called inflation. Through the inflation and the following phases, several key processes took place. One of them was that the elementary particles created in the Big Bang—with their masses and charges and other characteristics precisely de-signed for it to happen—took on new forms. From the "quark soup" of particles that resulted from the creation of mass in the universe, the quarks miraculously bunched together in threes to make protons and neutrons. These then gathered in little groups called nuclei, held together by the strong nuclear force, leading to the formation of atoms, once electrons, with their charges negative but equal in value to those of protons, began to orbit the nuclei through the action of the electromagnetic force. Hydrogen, in huge amounts, and some helium, and a small amount of lithium were thus immediately created.

Then the force of gravity took over and these elements bunched together to form the earliest stars and galaxies, and when the stars became dense enough they ignited in nuclear fires, aided by the action of the weak nuclear force. The stars proceeded to create heavier elements by the process of fusion. As stars lived out their lives, some died by shedding their element-rich atmospheres into space, while others exploded in supernovas, releasing heavier elements into the cosmos. The chemical elements created inside stars became the basic build-ing blocks of life. Throughout the process, the cosmological constant controlled the expansion of the universe so it would neither recollapse nor explode before life could evolve on at least one planet. Our sun is a later-generation star, and the disc that formed around it by mutual gravitational attraction of cosmic dust in the solar system created the Earth and the other planets, as matter made of hydrogen and heavier elements created by earlier stars coalesced 4.5 billion years ago. Our planet was thus endowed with the elements that would be needed to eventually bring about life: carbon, iron, nitrogen, oxygen, and so on.

The immensely complex configuration of the forces, masses, ratios, charges, and all other numerical specifications of the universe had to have been determined to such a staggeringly precise level that to assume it happened "by chance" or through the anthropic principle seems futile. Perhaps some great force of creation set the parameters for our existence to the exacting standards required for life to evolve. Chance alone is virtually impossible to have played a role, since, as noted earlier, the odds against a universe with life and intelligence are at most one to a number that has 1 followed by 10 raised to the power 117 zeros (based on the requirements of only one of the parameters!)— the odds are so staggeringly high against our existence that even talking about probability and chance in this context is unproductive.

The Earth, once it formed, went through geological and atmospheric changes and at some point, through a mystery we have not solved, very early life-forms emerged, including algae and other organisms that perform photosynthesis, which turned the carbon-dioxide- and nitrogen-rich atmosphere into one that has a significant amount of oxygen. Again by unknown or not well-understood processes, animal life emerged on Earth, and through evolution led to the emergence of higher and higher organisms.

After millions of years of evolution, human consciousness

emerged and a species that could think symbolically and logically and was able to create great art and science and mathematics and literature and language came into being. The emergence of consciousness and language and intelligence and symbolic thinking is not understood by science and we have so far not explaining satisfactorily how this miracle happened. The odds against the biological processes that created life in the first place, established the eukaryotic cells that make up advanced organisms, and led to intelligence and consciousness are immeasurably high. Evolution can tell us how animals and species move through time—how creatures advance up the ladder of life—but it does not explain the immensely improbable appearance of life and intelligence and consciousness. These mysteries remain unsolved.

The odds against a universe such as we have are fantastically great, and so are the odds against the emergence of life and the advent of intelligence. What we see around us is nothing less than a series of extremely unlikely events that could well be explained as a "miracle"—the miracle of the universe, the world, life, intelligence, and human beings contemplating the riddle of creation. In a sense, with the emergence of conscious beings on a planet orbiting an average yellow star on the outskirts of a spiral galaxy, out of billions, we call the Milky Way, within the immense vastness of space, surrounded by distant galaxies from which no signal of life or intelligence has ever been discovered, the entire universe of which we are part has gained a self-awareness and a consciousness.

We don't have good explanations for *everything* in science. Science is about learning about the world, and we have indeed made immense progress in understanding nature. Many of the misconceptions brought on by a literal interpretation of the Bible have been debunked by science: a young age for the Earth, its being the center of the universe, and the direct creation of species by God rather than their evolution from less advanced organisms. And we know that the sun doesn't stop its apparent motion through the sky for any reason, and that the earth rotates. But do these facts imply that God does not exist? Certainly not.

The New Atheists love to bring up the question: "If God made the universe, who made God?" It's a fair question to ask, but we obviously don't know the answer. And just because this question can't be answered doesn't mean that by asking it they somehow prove that God doesn't exist. It simply shows that God's existence and what, if anything, "created God" are well outside the realm of questions that science and mathematics can answer.

We don't fully understand what space is made of, and what the elements of physical space are and how they are stacked together. We don't know the level of infinity of the real line and whether the mathematical line has the properties of physical space. We don't know how space and time were created. We don't know what time really is. We don't know what caused the Big Bang. And we don't know who or what created God. What we do know is that the universe did not come out of the void all by itself: something preceded the Big Bang, and that "something" is unreachable to our science and may well remain so forever. We know that by some strange and mysterious mechanism all the constants of nature turned out to be exactly as they need to be for life to emerge, and the alternatives to a divine control that effected these incredibly unlikely conditions are no more likely than is the existence of God.

The same is true for the emergence of the molecules of life—the complicated DNA code-bearing helixes that propagate living systems on Earth—and the complex cells of living organisms that came into existence on this planet; and the evolution of beautiful, advanced life-forms; and finally the emergence of conscious, thinking, intelligent creatures that give the universe its self-awareness. The New Atheists have not shown us how all of these mysterious events in the history of the universe ever took place and how they could emerge and come together perfectly all by themselves without an external guiding force.

As we have seen, some truths are mathematically outside the reach of the human brain, despite its great abilities. As pure mathematics has shown us, some propositions cannot be proved or disproved from within a given system and require a way of getting "outside the box" in order to gain information about them that could lead to a proof or a disproof of any assertion. Our inherent inability to penetrate our cosmic past to a level that could get us back to the Big Bang and whatever caused it may be our main stumbling block on the way to understanding where we came from, why we came here, where we are going, and who or what made us. In any logical system there are un-provable assertions, and the question of God may well be one of these unknowns—forever to remain outside our reach.

If someday we should meet members of other, nonhuman civilizations, either through direct contact or through some form of radio communication, and be able to evaluate how advanced they are as compared to us, and learn about their belief systems and knowledge base, then perhaps we could learn more. But even so, going back to the Big Bang is outside the reach of any civilization, and it is therefore likely that even if such an extraordinary event should ever take place and we should learn about other creatures of the universe, we may still remain ignorant about the mystery of creation.

**RICHARD DAWKINS'S FINAL** line of attack against religion is the argument that the vast majority of prominent scientists are not religious. This kind of statement is deceptive. Many independent-minded people and intellectuals do not like the prescriptions and rituals of organized religion. And it's certainly true that religions are tradition-bound institutions that frequently have been antagonistic to change—both social and scientific. But this doesn't mean that many scientists don't see in nature and beyond it a force that is unknown to us and unknowable; that force can inspire in us a sense of humility and awe as well as the understanding that we don't know everything and may well never learn some important truths about the universe.

In 2009 I traveled to Jerusalem to interview one of the

greatest cosmologists working today: Jacob Bekenstein of the Hebrew University. In the 1970s, while he was a student of the renowned physicist John Wheeler at Princeton, it was Bekenstein who first carried out the calculation of the entropy of a black hole, showing that black holes have a nonzero temperature. Stephen Hawking mocked the discovery, and Bekenstein found himself the butt of jokes. But he was proven right, using mathematics, and later Hawking turned around completely and made a similar calculation, replicating Bekenstein's finding. The result is now called the Bekenstein-Hawking radiation law for black holes. As it happens, Bekenstein is also a sincerely religious man, one who follows the lifestyle of a modern Orthodox Jew. He finds absolutely no contradiction between his religious faith and his work in the frontiers of science to pursue the ultimate laws of the universe. So there are indeed leading scientists in the world today who happen to also be believers.

**IN THIS BOOK,** I have not proved the existence of God in any shape or form, and this has obviously not been my purpose. What I aimed to do was to argue—convincingly, I hope—that science has not disproved the existence of God. Since we do not know what God is and have no way of perceiving infinite power, infinite space, infinite time, infinite wisdom, infinite love, and other deep concepts we may associate with God, it is well outside the realm of the possible for us to ever hope to answer such questions. The God of literal interpretations of Scripture written for primitive peoples thousands of years ago certainly does not exist. And religions have their flaws, as all human institutions do. But God—a power well outside our ability to comprehend, transcending the creation of the universe we see around us—may well exist, and science has not, and will not, disprove it.

In so many ways, the same impulse to know the world and our place in it is at the roots of both science and spirituality. Both are attempts to illuminate the mysteries of our world and expand our vision of the greater whole. By charting the history of science, I hope these pages have shown how vital and awesome real science is. Throughout history, scientific discovery has brought us closer to the wonders of life and the universe—and immeasurably deepened our appreciation for creation. It engages the world and inspires the best in us. But the pursuit of truth should not be driven by zealous agenda. Nor should it overreach and speak with righteous authority where it's on unsolid ground. That's not science—and let's not allow those who falsely invoke its name to diminish us.

# Notes

## PROLOGUE

10   **"relatively close to the earth":** Harris, *Letter to a Christian Nation*, p. 68.

11   **"suffering and its alleviation":** Harris, *Letter*, p. 25.

12   **"psychotic delinquent":** Dawkins, *The God Delusion*, p. 59.

14   **"physics and cosmology as well":** Dawkins, *The God Delusion*, p. 143.

15   **"mass graves all over Bosnia":** Hitchens, *God Is Not Great*, p. 16.

20   **"remarkably similar entities":** James Wood, the *Guardian*, August 26, 2011.

21   **"( . . . at a site that has never been located)":** Christopher Hitchens, *God Is Not Great*, p. 117.

# CHAPTER 1

25   **images of the first dieties ever conceived:** Cohen, *La femme des origines*, p. 14.

27   **earliest religion in the human past:** Leroi-Gourhan, André, *Les religions de la préhistoire*; Leroi-Gourhan, André, *Le fil du temps*, pp. 201–40.

27   **although they remain undeciphered:** Tattersall, *Masters of the Planet*, pp. xiii–xiv.

27   **coast of Ecuador in the Americas:** Mann, *1491*, p. 21.

28   **exactly what was needed for agriculture to succeed:** Mithen, *After the Ice*, pp. 52–54.

28   **eleven to nine millennia before our time:** Cauvin, *Naissance des divinités*, pp. 43–55.

30   **"delicately modeled with plaster":** Mithen, *After the Ice*, p. 81.

32   **narrow waists and prominent buttocks:** Scarre, *Timelines*, p. 77.

32   **and an offering table:** Scarre, *Timelines*, p. 78.

35   **dating from about 6000 B.P.:** Israel Museum, Plate 20 in Dayagi-Mendels and Rozenberg, eds., *Chronicles of the Land*.

35   **a ram carries cornets:** Dayagi-Mendels and Rozenberg, eds., *Chronicles of the Land*, p. 31.

36   **one's protection and patronage:** Dayagi-Mendels and Rozenberg, eds., *Chronicles of the Land*, p. 31.

37   **"of the prescientific stage":** Jammer, *Concepts of Force*, p. 18.

37  **carry out the action of the force:** Jammer, *Concepts of Force*, p. 19.

39  **"to have great force":** Jammer, *Concepts of Force*, p. 21.

40  **the official religion of the empire:** Freemam, *A.D. 381*, p. 40.

40  **and a meaning to life:** Freeman, *A.D. 381*, pp. 40–41.

41  **there was simply no time:** Hawking and Mlodinow, *The Grand Design*, pp. 49–50.

46  **connection between mathematics and worship in Eastern religions:** Georges Coedes, "A Propos de l'Origine des Chiffres Arabes," *Bulletin of the School of Oriental and African Studies* 6 (1931): pp. 323–28.

48  **should we "cede to *religion* the right to tell us what is good and what is bad?":** Dawkins, *The God Delusion*, p. 80.

# CHAPTER 2

50  **"none of the religious myths has any truth to it, or in it":** Hitchens, *God Is Not Great*, p. 102.

51  **and six-meter-high wall:** *Science Daily*, February 2010, p. 1.

55  **chronology of the Hebrew kings described in the Old Testament:** Thiele, *The Mysterious Numbers of the Hebrew Kings*, pp. 69–78.

56  **in confirming some accounts of the Old Testament:** The inscription can be seen in the Israel Museum in Jerusalem.

57 **have been discovered and dated to this period:** Dayagi-Mendels, p. 74.

58 **the rebuilding of the Second Temple by King Herod:** Display in the Israel Museum.

59 **has been discovered:** Display in the Israel Museum.


## CHAPTER 3


69 **Andreas Osiander:** Gingerich, *The Book Nobody Read.*

70 **was sympathetic to him:** Sobel, *Galileo's Daughter.*

74 **this deepening dispute:** Peter Barker and Bernard R. Goldstein, "Theological Foundations of Kepler's Astronomy," *Osiris* 16 (2001): pp. 88–113.

75 **until the end of his life:** See Gaukroger.

77 **at any given time:** Mehl, *Descartes,* 2001.

78 **Hobbes, Bacon, and Galileo:** Robert, *Leibniz: Vie et Oeuvre,* p. 11.

79 **basic Greek element of space:** Pierre Cartier, "A Mad Day's Work: From Grothendieck to Connes and Kontsevich The Evolution of Concepts of Space and Symmetry," *Bulletin of the American Mathematical Society* 38, no. 4 (2001): pp. 389–408.

## CHAPTER 4

85    **"( . . . it explains many things)":** Boyer and Merzbach, *A History of Mathematics*, p. 494.

90    **"incessant motion through space":** Foucault's private notebook, quoted in Aczel, *Pendulum*, p. 3.

91    **Foucault's proof that the Earth, indeed, rotates:** See Aczel, *Pendulum*, p. 55.

## CHAPTER 5

104   **speaks volumes about his attitude to God:** Calaprice, *Dear Professor Einstein*, p. 30.

105   **"whether God [sic] had any choice in the creation of the universe":** Krauss, *A Universe from Nothing*, p. 160.

105   **"labeled, motivated by Spinoza, with the moniker 'God' ":** Krauss, *A Universe from Nothing*, p. 160.

106   **"Science without religion is lame, religion without science is blind":** Dawkins, *The God Delusion*, p. 36.

106   **"never denied this but have expressed it clearly":** Dawkins, *The God Delusion*, p. 36.

106   **"become a member of the Jewish religious community":** Frank, *Einstein: His Life and Times*, p. 83.

106   **"became particularly friendly with Hugo Bergmann and Max Brod":** Frank, *Einstein: His Life and Times*, p. 84.

107 **as compared with the Jewish intellectuals of that city:** Fölsing, *Albert Einstein*, p. 279.

107 **"I discovered that I was a Jew":** Fölsing, *Albert Einstein*, p. 488.

107 **"that substantial portions of our spiritual inheritance may have passed down to me":** Fölsing, *Albert Einstein*, p. 490.

## CHAPTER 6

112 **"we all, literally, emerged from quantum nothingness":** Krauss, *A Universe from Nothing*, p. 98.

117 **"has no single past, or history":** Hawking and Mlodinow, *The Grand Design*, p. 82.

119 **" 'there is a lot of material to cover in this course' ":** Zee, *Quantum Field Theory in a Nutshell*, pp. 7–9.

120 **"then the screen is not really there. Very Zen!":** Zee, *Quantum Field Theory in a Nutshell*, p. 9.

120 **"of the great structures of 'the Real' ":** d'Espagnat, *On Physics and Philosophy*, p. 455.

121 **"the idea of a 'something' that we cannot conceptualize":** d'Espagnat, *On Physics and Philosophy*, pp. 455–56.

122 **"the time of empirical reality":** d'Espagnat, *On Physics and Philosophy*, p. 457.

123 **"It is just a mathematical convenience":** John Bell, *Speakable*, p. 53.

123 **"i.e., between one set of beables and another"**: John Bell, *Speakable*, p. 53.

## CHAPTER 7

128 **"Quantum Origin of the Universe"**: Alexander Vilenkin, "Quantum Origin of the Universe," *Nuclear Physics B* 252 (1985): pp. 141–52.

128 **"tested by overall consistency of our picture of the universe"**: Alexander Vilenkin, "Quantum Origin of the Universe" *Nuclear Physics B* 252 (1985): p. 152.

129 **"where by 'nothing' I mean a space with no classical space time"**: Alexander Vilenkin, "Quantum Origin of the Universe" *Nuclear Physics B* 252 (1985): p. 141.

133 **"Is the Universe a Vacuum Fluctuation?"**: Edward P. Tryon, "Is the Universe a Vacuum Fluctuation?" *Nature* 246 (December 14, 1973): pp. 396–97.

133 **proposes the idea that our universe may have a sum of zero for all its *quantum numbers***: Edward P. Tryon, "Is the Universe a Vacuum Fluctuation?" *Nature* 246 (December 14, 1973): pp. 396–97.

## CHAPTER 8

138 **Brian Greene:** Greene, *The Hidden Reality: Parallel Universes and the Deep Laws of the Cosmos.*

146 **"consciousness raised by natural selection":** Dawkins, *Delusion*, p. 175.

## CHAPTER 9

152 **represented in the figure below:** Penrose, *The Road to Reality*, p. 18.

153 **"not rooted in physical structures":** Penrose, *The Road to Reality*, p. 20.

154 **between mathematics and physics and related sciences:** Eugene Wigner, "The Unreasonable Effectiveness of Mathematics in the Natural Sciences," *Communications on Pure and Applied Mathematics* 13 (1960): pp. 1–14.

155 **"have to do with a circle":** Wigner, "The Unreasonable Effectiveness of Mathematics in the Natural Sciences," p. 1.

156 **"being ruled out by the laws of probability":** Dawkins, *The God Delusion*, p. 68.

156 **"done this at one time or another":** Dawkins, *The God Delusion*, p. 72.

156 **"50 percent starting likelihood each":** Dawkins, *The God Delusion*, p. 133.

157 **British statistician Harold Jeffreys:** Jeffreys, *Theory of Probability.*

158 **that arises from the use of real data:** Box and Tiao, *Bayesian Inference in Statistical Analysis*, p. 41.

159 **"can be drawn from the data":** Box and Tiao, *Bayesian Inference in Statistical Analysis*, p. 58.

160 **"inroads into the territory of agnosticism":** Dawkins, *The God Delusion*, p. 96.

162 **"(a good figure for this kind of study)":** Dawkins, *The God Delusion*, p. 128.

164 **in a modern statistical hypothesis-testing procedure:** Stigler, *The History of Statistics: The Measurement of Uncertainty Before 1900.*

## CHAPTER 10

169 **René Thom derived a theory about abrupt, unforeseen change:** David Aubin, "Forms of Explanations of the Catastrophe Theory of René Thom: Topology, Morphogenesis, and Structuralism," in *Growing Explanations: Historical Perspective on the Sciences of Complexity.* M. N. Wise, ed., pp. 95–130.

172 **Highly nonlinear mathematical systems are notoriously volatile:** Kautz, *Chaos.*

174 **movements from point to point on the surface of a fractal are bizarre and chaotic:** See, for example, Mandelbrot, *Fractals and Chaos.*

174 **can cause a huge, unexpected change at a distant location:** See, for example, Edward Norton Lorentz, "Deterministic Non-periodic Flow," *Journal of the Atmospheric Sciences* 20, no. 2 (1963): pp. 130–41.

## CHAPTER 11

178 **Penrose writes, in *The Road to Reality*:** Roger Penrose, *The Road to Reality*, pp. 762–65.

179 **equal to that in the observable universe:** Roger Penrose, *The Road to Reality*, p. 762.

180 **"we can get with the second possibility":** Roger Penrose, *The Road to Reality*, pp. 764–65.

181 **comes "from the revelation of God":** Reported in Ferguson, *Stephen Hawking: An Unfettered Mind*, p. 102.

181 **"start to discuss the origins of the universe":** Boslough, *Beyond the Black Hole*, p. 100.

182 **"a universe that has produced life like ours are immense":** Boslough, *Beyond the Black Hole*, p. 105.

182 **"we wouldn't be around to ask the question":** Ferguson, *Stephen Hawking*, p. 104.

183 **"grants it that luck":** Dawkins, *The God Delusion*, p. 169.

183 **"my consciousness has been raised by Darwin":** Dawkins, *The God Delusion*, footnote, p. 173.

185 **"'we don't know how he pushed his pencil'":** Feynman, *QED: The Strange Theory of Light and Matter*, p. 131.

188   **This is certainly one interesting view of the universe.**
Richard Dawkins seems eager to exploit everything that
comes his way, including the good names of famous scientists. Here is how he exploited Steven Weinberg in his
book *The God Delusion*:

As the Nobel Prize–winning American physicist Steven
Weinberg said, "Religion is an insult to human dignity.
With or without it, you'd have good people doing good
things and evil people doing evil things. But for good
people to do evil things, it takes a religion." (Dawkins,
*The God Delusion*, p. 283.)

It is true that Steven Weinberg said this, but—atheist
as he may be—he also once said, at a conference in 2006
attended by Richard Dawkins, that much as he would
like to see civilization emerge from the tyranny of religion, when it happens, "I think we will miss it, like a
crazy old aunt who tells lies and causes us all kinds of
trouble, but was beautiful once and was with us a long
time." The combative Dawkins interjected: "I won't
miss her at all!" (reported in *Newsweek*, Nov. 10, 2006).
When I met him at the Puebla conference in November 2010, I told Dawkins that I felt that Steven Weinberg, who is also an atheist, does not hold his extreme
views. Dawkins jumped up and said: "Yes he does! Weinberg holds exactly my views on religion." "I don't think
so," I answered, "he does not agree with you about the

'crazy aunt.'" "Yes," he admitted grudgingly, and walked away.

## CHAPTER 12

196  **but in a more passive form:** George Levine, Introduction to Darwin, *On the Origin of Species*, p. xviii.

197  **"and are being, evolved":** Darwin, *On the Origin of Species*, p. 384.

198  **"or by the doctrine of final causes":** Gould, *The Structure of Evolutionary Theory*, p. 100; quoting Darwin's *On the Origin of Species*, p. 435.

198  **into the ultimate origins of things:** Gould, *The Structure of Evolutionary Theory*, p. 101.

199  **"even make *probability* judgments on the question":** Dawkins, *The God Delusion*, p. 81.

199  **"like evolution, paleontology, and baseball":** Gould, *Rocks of Ages*, p. 3

203  **explained by endowing evolutionary advantages:** Wilson, *The Social Conquest of Earth*, pp. 109–20.

## CHAPTER 13

212  **around 7 million years ago:** Walker and Shipman, *The Wisdom of the Bones*, p. 150.

216 **"to blur the moral principle and destroy its absolute-ness"**: Dawkins, *The God Delusion*, p. 340.

219 **"no wonder consciousness is still a mystery"**: Dennett, *Consciousness Explained*, p. 255.

219 **"They were not developed to perform particularly human actions"**: Dennett, *Consciousness Explained*, pp. 253–54.

221 **"are among the commonplaces of engineering science"**: Dawkins, *The Selfish Gene*, p. 50.

221 **"by lists of electronically coded numbers"**: Dawkins, *The Selfish Gene*, p. 58.

222 **separates us from all our predecessors, ancestors, and other animals**: Tattersall, *Becoming Human*, p. xx.

## CHAPTER 14

237 **"investigate metamathematical questions"**: Dawson, *Logical Dilemmas*, p. 262.

238 **"of the sort he had envisioned"**: Dawson, *Logical Dilemmas*, p. 263.

239 **"that the prejudice is a mistake"**: Hao Wang, *A Logical Journey*, p. 83.

## CHAPTER 15

242  **the humiliated Diderot withdrew and, the next day, packed his bags and returned to France:** See R. J. Gillings, "The So-Called Euler-Diderot Incident," *American Mathematical Monthly* 61, no. 2 (February 1954): pp. 77–80. There are several different versions of this story. The reference above implies that the only accurate version is that reported by D. Thiebault in a memoir published in Paris in 1804.

251  **that the vast majority of prominent scientists are not religious:** Dawkins, *The God Delusion*, p. 68.

# Bibliography

*This list contains books that are quoted in this work or were consulted by the author in its preparation. Scientific articles, lectures, and other less-accessible sources of information appear only in the notes.*

Aczel, Amir D. *God's Equation: Einstein, Relativity, and the Expanding Universe.* New York: Basic Books, 1999.

———. *The Mystery of the Aleph: Mathematics, Kabbalah, and the Search for Infinity.* New York: Basic Books, 2000.

———. *Entanglement: The Greatest Mystery in Physics.* New York: Basic Books, 2002.

———. *Pendulum: Léon Foucault and the Triumph of Science.* New York: Atria, 2003.

———. *Descartes's Secret Notebook.* New York: Broadway, 2005.

———. *The Cave and the Cathedral: How a Real-Life Indiana Jones and a Renegade Scholar Decoded the Ancient Art of Man.* New York: Wiley, 2009.

———. *A Strange Wilderness: The Lives of the Great Mathematicians.* New York: Sterling, 2012.

Arsuaga, Juan Luis. *The Neanderthal's Necklace: In Search of the First Thinkers*. Translated by Andy Klatt. New York: Four Walls Eight Windows, 2002.

Barrow, John D. *Pi in the Sky: Counting, Thinking, and Being*. New York: Little, Brown, 1992.

De Beaune, Sophie A. *Les hommes au temps de Lascaux*. Paris: Hachette, 1995.

Bell, Eric Temple. *Men of Mathematics*. New York: Simon and Schuster, 1937.

Bell, John S. *Speakable and Unspeakable in Quantum Mechanics*. New York: Cambridge University Press, 1993.

Berra, Tim M. *Evolution and the Myth of Creationism*. Stanford, CA: Stanford University Press, 1990.

Boaz, Noel T., and Russell L. Ciochon. *Dragon Bone Hill: An Ice-Age Saga of Homo Erectus*. New York: Oxford University Press, 2004.

Bohm, David. *Causality and Chance in Modern Physics*. Foreword by Louis de Broglie. Philadelphia: University of Pennsylvania Press, 1957.

———. *Quantum Theory*. New York: Dover, 1989.

Bonola, Roberto. *Non-Euclidean Geometry*. New York: Dover, 1955.

Born, Max. *Einstein's Theory of Relativity*. New York: Dover, 1962.

Boslough, John. *Beyond the Black Hole: Stephen Hawking's Universe*. Glasgow, UK: Fontana/Collins, 1984.

De Botton, Alain. *Religion for Atheists: A Non-Believer's Guide to the Uses of Religion*. New York: Pantheon, 2012.

Box, George E. P., and George C. Tiao. *Bayesian Inference in Statistical Analysis*. Reading, MA: Addison-Wesley, 1973.

Boyer, Carl, and Uta Merzbach. *A History of Mathematics*. Second Edition. New York: Wiley, 1993.

Calaprice, Alice. *Dear Professor Einstein: Albert Einstein's Letters to and from Children*. New York: Prometheus, 2002.

Carroll, James. *Constantine's Sword: The Church and the Jews—A History*. New York: Mariner, 2001.

Cauvin, Jacques. *Naissance des divinités, naissance de l'agriculture*. Paris: Flammarion, 1997.

Charraud, Nathalie. *Infini et inconscient: Essai sur Georg Cantor*. Paris: Anthropos, 1994.

Ciufolini, Ignazio, and John Archibald Wheeler. *Gravitation and Inertia*. Princeton, NJ: Princeton University Press, 1995.

Clark, Ronald W. *Einstein: The Life and Times*. New York: Avon, 1972.

Close, Frank. *The Infinity Puzzle*. New York: Basic Books, 2011.

Cohen, Claudine. *La femme des origines: Images de la femme dans la préhistoire occidentale*. Paris: Belin-Herscher, 2003.

Cole, K. C. *The Universe and the Teacup: The Mathematics of Truth and Beauty*. New York: Harcourt Brace, 1998.

Collins, Francis S. *The Language of God: A Scientist Presents Evidence for Belief*. New York: Free Press, 2007.

Croswell, Ken. *Planet Quest: The Epic Discovery of Alien Solar Systems*. New York: Free Press, 1997.

Darwin, Charles. *On the Origin of Species*. New York: Barnes and Noble, 2004 (reprint of the 1859 edition).

David, F. N. *Games, Gods, and Gambling: A History of Probability and Statistical Ideas.* New York: Dover, 1998.

Davies, Paul. *The Mind of God.* New York: Simon and Schuster, 1991.

———. *About Time: Einstein's Unfinished Revolution.* New York: Simon and Schuster, 1995.

Dawkins, Richard. *The God Delusion.* New York: Houghton Mifflin Harcourt, 2006.

———. *The Selfish Gene.* New York: Oxford University Press, 1989.

Dawson, John W., Jr. *Logical Dilemmas: The Life and Work of Kurt Gödel.* Wellesley, MA: A. K. Peters, 1997.

Dayagi-Mendels, M., and S. Rozenberg, eds. *Chronicles of the Land: Archaeology in the Israel Museum, Jerusalem.* Jerusalem: The Israel Museum, 2011.

Dennett, Daniel. *Consciousness Explained.* New York: Little, Brown, 1991.

———. *Darwin's Dangerous Idea.* New York: Simon and Schuster, 1995.

Descartes, René. *Oeuvres philosophiques.* Volume I: 1618–1637. Paris: Garnier, 1997.

Dirac, Paul A. M. *The Principles of Quantum Mechanics.* Fourth Edition. New York: Oxford University Press, 1958.

Einstein, Albert. *Relativity: The Special and the General Theory.* New York: Crown, 1961.

Einstein, Albert, A. Lorentz, Hermann Weil, and H. Minkowski. *The Principle of Relativity.* New York: Dover, 1952.

d'Espagnat, Bernard. *On Physics and Philosophy*. Princeton, NJ: Princeton University Press, 2006.

Ferguson, Kitty. *Stephen Hawking: An Unfettered Mind*. New York: Palgrave/Macmillan, 2012.

Feynman, Richard P. *QED: The Strange Theory of Light and Matter*. Princeton, NJ: Princeton University Press, 1985.

———. *The Character of Physical Law*. Cambridge, MA: MIT Press, 2001.

———. *Six Not-So-Easy Pieces*. New York: Basic Books, 2011.

Fölsing, Albrecht. *Albert Einstein*. New York: Penguin, 1998.

Frank, Philipp. *Einstein: His Life and Times*. Translated by George Rosen. New York: Da Capo, 1989.

Freeman, Charles. *A.D. 381: Heretics, Pagans, and the Dawn of the Monotheistic State*. New York: Overlook, 2008

French, A. P., and Edwin F. Taylor. *An Introduction to Quantum Physics*. New York: Norton, 1978.

Freund, Jürgen. *Special Relativity for Beginners*. Singapore: World Scientific, 2008.

Gapaillard, Jacques. *Et pourtant elle tourne!: Le mouvement de la Terre*. Paris: Seuil, 1993.

Geroch, Robert. *Mathematical Physics*. Chicago, IL: The University of Chicago Press, 1985.

Gilmore, Robert. *Lie Groups, Lie Algebras, and Some of Their Applications*. New York: Dover, 2002.

Gingerich, Owen. *The Book Nobody Read: Chasing the Revolutions of Nicolaus Copernicus*. New York: Walker, 2004.

Gödel, Kurt. *Collected Works.* Volume II, Publications 1938–1974. Edited by Solomon Feferman, et al. New York: Oxford University Press, 1990.

Goldsmith, Donald. *Einstein's Greatest Blunder? The Cosmological Constant and Other Fudge Factors in the Physics of the Universe.* Cambridge, MA: Harvard University Press, 1995.

Gondhalekar, Prabhakar. *The Grip of Gravity: The Quest to Understand the Laws of Motion and Gravitation.* New York: Cambridge University Press, 2001.

Gould, Stephen Jay. *The Structure of Evolutionary Theory.* Cambridge, MA: Belknap Harvard, 2002.

———. *Rocks of Ages: Science and Religion in the Fullness of Life.* New York: Ballantine, 1999.

———. *Ever Since Darwin: Reflections in Natural History.* New York: Norton, 1992.

Greene, Brian. *The Fabric of the Cosmos: Space, Time, and the Texture of Reality.* New York: Knopf, 2003.

———. *The Hidden Reality: Parallel Universes and the Deep Laws of the Cosmos.* New York: Knopf, 2011.

Guth, Alan. *The Inflationary Universe.* New York: Basic Books, 1998.

Hacking, Ian. *The Emergence of Probability: A Philosophical Study of Early Ideas about Probability, Induction, and Statistical Inference.* Second Edition. New York: Cambridge University Press, 2006.

Hajnal, András, and Peter Hamburger. *Set Theory.* New York: Cambridge University Press, 1999.

Harris, Sam. *The End of Faith: Religion, Terror, and the Future of Reason*. New York: Norton, 2004.

———. *Letter to a Christian Nation*. New York: Knopf, 2006.

———. *The Moral Landscape: How Science Can Determine Human Values*. New York: Free Press, 2010.

Hawking, Stephen. *The Illustrated A Brief History of Time* and *The Universe in a Nutshell*. New York: Bantam, 2001.

Hawking, Stephen, and G. F. R. Ellis. *The Large Scale Structure of Space-Time*. New York: Cambridge University Press, 1973.

Hawking, Stephen, and Leonard Mlodinow. *The Grand Design*. New York: Bantam, 2010.

Heilbron, J. L. *The Sun in the Church: Cathedrals as Solar Observatories*. Cambridge, MA: Harvard University Press, 1999.

Heisenberg, Werner. *Ordnung der Wirklichkeit*. Munich: R. Piper, 1989.

Hitchens, Christopher. *The Portable Atheist*. New York: Perseus, 2007.

———. *God Is Not Great: How Religion Poisons Everything*. New York: Twelve, 2009.

Holton, Gerald. *Thematic Origins of Scientific Thought: Kepler to Einstein*. Cambridge, MA: Harvard University Press, 1973.

Hooper, Dan. *Nature's Blueprint: Supersymmetry and the Search for a Unified Theory of Matter and Force*. New York: HarperCollins, 2008.

Hoskin, Michael, ed. *The Cambridge Illustrated History of Astronomy*. New York: Cambridge University Press, 1997.

Huff, Darrell. *How to Lie with Statistics*. New York: Norton, 1954.

Isaacson, Walter. *Einstein: His Life and Universe*. New York: Simon and Schuster, 2008.

Jammer, Max. *Concepts of Force*. Cambridge, MA: Harvard University Press, 1957. Reissued, New York: Dover, 1999.

Jeffreys, Harold. *Theory of Probability*. New York: Oxford University Press, 1998 (reissue of 1939 edition).

Kaku, Michio. *Hyperspace: A Scientific Odyssey Through Parallel Universes, Time Warps, and the 10th Dimension*. New York: Oxford University Press, 1994.

———. *Parallel Worlds: A Journey Through Creation, Higher Dimensions, and the Future of the Cosmos*. New York: Anchor, 2006.

Kandel, Eric R. *The Age of Insight: The Quest to Understand the Unconscious in Art, Mind, and Brain, from Vienna 1900 to the Present*. New York: Random House, 2012.

Kane, Gordon, and Aaron Pierce, eds. *Perspectives on LHC Physics*. Hackensack, NJ: World Scientific, 2008.

Kant, Immanuel. *Critique of Pure Reason*. New York: Dover, 2003.

Kautz, Richard. *Chaos: The Science of Predictable Random Motion*. Oxford, UK: Oxford University Press, 2011.

Kelley, John L. *General Topology*. New York: Van Nostrand, 1955.

Kolata, Gina. *Clone: The Road to Dolly and the Path Ahead*. New York: Morrow, 1998.

Kosmann-Schwarzbach, Yvette. *The Noether Theorems: Invariance and Conservation Laws in the Twentieth Century*. Translated by Bertram E. Schwarzbach. New York: Springer, 2011.

Krauss, Lawrence. *A Universe from Nothing: Why There Is Something Rather than Nothing*. New York: Free Press, 2012.

Kurzweil, Ray. *The Singularity Is Near*. New York: Penguin, 2006.

Lai, C. H., ed. *Gauge Theory of Weak and Electromagnetic Interactions: Selected Papers*. Singapore: World Scientific, 1981.

Lehrer, Jonah. *How We Decide*. New York: Mariner, 2010.

Leroi-Gourhan, André. *Les religions de la préistoire*. Paris: Quadrige/PUF, 1964.

———. *Le fil du temps*. Paris: Fayard, 1983.

Levy, Silvio, ed. *Flavors of Geometry*. New York: Cambridge University Press, 1997.

Livio, Mario. *Is God a Mathematician?* New York: Simon and Schuster, 2009.

Maimonides, Moses. *The Guide for the Perplexed*. Translated by M. Friedlander. New York: Dover, 1956.

Majid, Shahn, ed. *On Space and Time*. New York: Cambridge University Press, 2008.

Mandelbrot, Benoit. *Fractals and Chaos*. Berlin: Springer Verlag, 2004.

Mann, Charles C. *1491: New Revelations of the Americas Before Columbus*. New York: Vintage, 2006.

Mehl, Edouard. *Descartes en Allemagne 1619-1620*. Strasbourg, France: Presses Universitaires de Strasbourg, 2001.

Messiah, Albert. *Quantum Mechanics*. Volumes I and II. New York: Dover, 1999.

Miller, Arthur I. *Deciphering the Cosmic Number: The Strange Friendship of Wolfgang Pauli and Carl Jung*. New York: Norton, 2009.

Mithen, Steven. *After the Ice: A Global Human History 20,000–5,000 B.C.* Cambridge, MA: Harvard University Press, 2006.

Mohen, J. P., and Y. Taborin. *Les sociétés de la Préhistoire*. Paris: Hachette, 2005.

Nambu, Y. *Quarks: Frontiers in Elementary Particle Physics*. Philadelphia: World Scientific, 1985.

Neugebauer, Otto. *The Exact Sciences in Antiquity*. New York: Dover, 1969.

Oerter, Robert. *The Theory of Almost Everything: The Standard Model, the Unsung Triumph of Modern Physics*. New York: Plume, 2006.

Pais, Abraham. *Niels Bohr's Times: In Physics, Philosophy, and Polity*. New York: Oxford University Press, 1991.

———. *Subtle Is the Lord: The Science and the Life of Albert Einstein*. New York: Oxford University Press, 2005.

Pasachoff, Jay. *Astronomy: From the Earth to the Universe*. Fifth edition. New York: Saunders, 1998.

Penrose, Roger. *The Emperor's New Mind*. New York: Oxford University Press, 1989.

———. *The Road to Reality: A Complete Guide to the Laws of the Universe*. New York: Knopf, 2005.

Pinker, Steven. *The Stuff of Thought: Language as a Window into Human Nature.* New York: Viking, 2007.

Poincaré, Henri. *The Value of Science.* New York: Modern Library, 2001.

Rabinovitch, Nahum L. *Probability and Statistical Inference in Ancient and Medieval Jewish Literature.* Toronto: University of Toronto Press, 1973.

Reichenbach, Hans. *The Philosophy of Space and Time.* Translated by Maria Reichenbach. New York: Dover, 1958.

Robert, Jean-Michel. *Leibniz: Vie et Oeuvre.* Paris: Pocket, 2003

Rodis-Lewis, Geneviève. *Descartes: Biographie.* Paris: Calmann-Lévy, 1995.

Ross, Sheldon M. *Introduction to Probability Models.* New York: Academic Press, 1972.

du Sautoy, Marcus. *Symmetry: A Journey into the Patterns of Nature.* New York: HarperCollins, 2008.

Scarre, Chris, ed. *Smithsonian Timelines of the Ancient World.* New York: Dorling Kindersley, 1993.

Schilpp, Paul Arthur, ed. *Albert Einstein: Philosopher-Scientist.* New York: MJF Books, 1970.

Schuster, Heinz Georg. *Deterministic Chaos: An Introduction.* New York: Wiley, 1987.

Smoot, George. *Wrinkles in Time: Witness to the Birth of the Universe.* Harper Perennial, 2007.

Sobel, Dava. *Galileo's Daughter: A Historical Memoir of Science, Faith, and Love.* New York: Walker, 2000.

Solecki, Ralph S. *Shanidar: The First Flower People.* New York: Knopf, 1971.

Stigler, Stephen M. *The History of Statistics: The Measurement of Uncertainty Before 1900.* Cambridge, MA: Belknap Harvard, 1990.

Stoker, J. J. *Differential Geometry.* New York: Wiley, 1969.

Susskind, Leonard. *The Black Hole War: My Battle with Stephen Hawking to Make the World Safe for Quantum Mechanics.* New York: Little, Brown, 2008.

Tattersall, Ian. *Becoming Human: Evolution and Human Uniqueness.* New York: Harcourt Brace, 1998.

———. *The Monkey in the Mirror: Essays on the Science of What Makes Us Human.* New York: Harcourt, 2002.

———. *Masters of the Planet: The Search for Our Human Origins.* New York: Palgrave/Macmillan, 2012.

Teilhard de Chardin, Pierre. *The Phenomenon of Man.* New York: Harper Perennial Modern Classics, 2008.

———. *Lettres de voyage.* Paris: Grasset, 1956.

Thiele, Edwin R. *The Mysterious Numbers of the Hebrew Kings.* Grand Rapids, MI: Kregel, 1983.

Vilenkin, Alex. *Many Worlds in One: The Search for Other Universes.* New York: Hill and Wang, 2007.

Walker, Alan, and Pat Shipman. *The Wisdom of the Bones: In Search of Human Origins.* New York: Knopf, 1996.

Wang, Hao. *A Logical Journey: From Gödel to Philosophy.* Cambridge, MA: MIT Press, 1996.

Weinberg, Steven. *Gravitation and Cosmology: Principles and Applications of the General Theory of Relativity*. New York: Wiley, 1972.

————. *The Quantum Theory of Fields*. Volumes I, II, and III. New York: Cambridge University Press, 2005.

Weyl, Hermann. *The Theory of Groups and Quantum Mechanics*. New York: Dover, 1931.

Wick, David. *The Infamous Boundary: Seven Decades of Heresy in Quantum Physics*. New York: Copernicus, 1996.

Wickham, Chris. *The Inheritance of Rome: Illuminating the Dark Ages 400-1000*. New York: Penguin, 2010.

Wilczek, Frank. *The Lightness of Being: Mass, Ether, and the Unification of Forces*. New York: Basic Books, 2008.

Wilson, E. O. *Consilience: The Unity of Knowledge*. New York: Vintage, 1999.

————. *The Social Conquest of Earth*. New York: Liveright, 2012.

Winchester, Simon. *The Map That Changed the World*. New York: HarperCollins, 2001.

Wise, M. N., ed. *Growing Explanations: Historical Perspective on the Science of Complexity*. Durham, NC: Duke University Press, 2004.

Zee, A. *Quantum Field Theory in a Nutshell*. Princeton, NJ: Princeton University Press, 2003.

Zwirn, Hervé. *Les limites de la connaissance*. Paris: Odile Jacob, 2000.

# Index

Page numbers in *italics* refer to illustrations.

*Photo by Debra Gross Aczel*

## About the Author

**Amir D. Aczel, Ph.D.,** received graduate degrees in mathematics from the University of California at Berkeley and the University of Oregon. He is the author of the acclaimed *Fermat's Last Theorem,* which has been published in twenty-eight languages and was nominated for a Los Angeles Times Book Prize, and many other works of nonfiction. In 2012, he was awarded a Sloan Foundation grant for his groundbreaking research on the origin of numbers; in 2004, he was awarded the prestigious John Simon Guggenheim Memorial Foundation Fellowship. From 2005 to 2007, Aczel was a visiting scholar at Harvard University. He is currently a research fellow in the history of science at Boston University. He also writes for *Discover* magazine, regularly publishes in *Scientific American,* and has written science pieces for the *New York Times* and *Wall Street Journal.*